BIOLOGY

Answer Key & Parent Companion

Science
Shepherd

SCIENCE SHEPHERD
BIOLOGY ANSWER KEY & PARENT COMPANION

Published by:
Ohana Life Press, LLC
1405 Capitol Dr.
Suite C-202
Pewaukee, WI 53072
www.ScienceShepherd.com

Written by Scott Hardin, MD

ACKNOWLEDGEMENTS:

Cover Design
Alex Hardin

Photo Credits
iStock.com

Copy Editing
Stacie O'Brien

Graphic Design
Jason Brown

ISBN: 978-0-9814587-2-4

Table of Contents

Introduction

The purpose of this booklet is two-fold; first, it contains the answers to all of the test and study questions that you (or your student) can use to grade tests and check progress on each chapter. Second, the Parent Companion section provides brief summaries of every chapter's sub-sections in the textbook. The summarized information is designed to give you rapid access to the main concepts your student is studying. In addition, you will also find one or two questions with answers that you can ask your student, if you are so inclined. Even if you know nothing about Biology, since the answers are also provided with the questions, you can use this section to ask specific questions of your student to see how they are progressing and grasping new concepts. Many parents use the summarized material and questions in the Companion to help them stay connected to what their child is learning by asking them one or more questions every now and then. However, it is totally optional and other parents do not use this section at all. Please take a little time and familiarize yourself with the Companion to see if it will fit into your homeschool!

Schedule

WEEK 1	DAY ONE	DAY TWO	DAY THREE	DAY FOUR	DAY FIVE
TEXTBOOK	Sections 1.0-1.9	Sections 1.10-1.13	Sections 1.14-1.17; chapter questions	Catch up/ study day	Sections 2.0-2.14
LAB MANUAL		Lab 1, or	Lab 1		
TEST BOOKLET					

WEEK 2	DAY ONE	DAY TWO	DAY THREE	DAY FOUR	DAY FIVE
TEXTBOOK	Sections 2.15-2.21	Sections 2.22-2.28; chapter questions	Catch up/ study day	Sections 3.0-3.8	Sections 3.9-3.15
LAB MANUAL					
TEST BOOKLET					

WEEK 3	DAY ONE	DAY TWO	DAY THREE	DAY FOUR	DAY FIVE
TEXTBOOK	Sections 3.16-3.19	Sections 3.20-3.26	Sections 3.27-3.33; chapter questions	Catch up/ study day	
LAB MANUAL					
TEST BOOKLET					Test #1

WEEK 4	DAY ONE	DAY TWO	DAY THREE	DAY FOUR	DAY FIVE
TEXTBOOK	Sections 4.0-4.15	Sections 4.16-4.22	Sections 4.23-4.27	Sections 4.28-4.33; chapter questions	Catch up/ study day
LAB MANUAL			Lab 2, or	Lab 2	
TEST BOOKLET					

WEEK 5	DAY ONE	DAY TWO	DAY THREE	DAY FOUR	DAY FIVE
TEXTBOOK	Sections 5.0-5.6	Sections 5.7-5.17	Sections 5.18-5.20; chapter questions	Catch up/ study day	
LAB MANUAL			Lab 3		
TEST BOOKLET					Test #2

WEEK 6	DAY ONE	DAY TWO	DAY THREE	DAY FOUR	DAY FIVE
TEXTBOOK	Sections 6.0-6.9	Sections 6.10-6.17	Sections 6.18-6.22; chapter questions	Catch up/ study day	Sections 7.0-7.8
LAB MANUAL					
TEST BOOKLET					

WEEK 7	DAY ONE	DAY TWO	DAY THREE	DAY FOUR	DAY FIVE
TEXTBOOK	Sections 7.9-7.15	Sections 7.16-7.19	Sections 7.20-7.23; chapter questions	Catch up/ study day	Sections 8.0-8.6
LAB MANUAL					
TEST BOOKLET					

WEEK 8	DAY ONE	DAY TWO	DAY THREE	DAY FOUR	DAY FIVE
TEXTBOOK	Sections 8.7-8.10	Sections 8.11-8.14	Sections 8.15-8.19; chapter questions	Catch up/ study day	
LAB MANUAL					
TEST BOOKLET					Test #3

WEEK 9	DAY ONE	DAY TWO	DAY THREE	DAY FOUR	DAY FIVE
TEXTBOOK	Sections 9.0-9.4	Sections 9.5-9.9	Sections 9.10-9.18	Sections 9.19-9.22; chapter questions	Catch up/ study day
LAB MANUAL					
TEST BOOKLET					

WEEK 10	DAY ONE	DAY TWO	DAY THREE	DAY FOUR	DAY FIVE
TEXTBOOK	Sections 10.0-10.4	Sections 10.5-10.8	Sections 10.9-10.14	Sections 10.15-10.18; chapter questions	Catch up/ study day
LAB MANUAL			Lab 4, or	Lab 4	
TEST BOOKLET					

WEEK 11	DAY ONE	DAY TWO	DAY THREE	DAY FOUR	DAY FIVE
TEXTBOOK	Sections 11.0-11.5	Sections 11.6-11.9	Sections 11.10-11.15	Sections 11.16-11.20; chapter questions	Catch up/ study day
LAB MANUAL					
TEST BOOKLET					

WEEK 12	DAY ONE	DAY TWO	DAY THREE	DAY FOUR	DAY FIVE
TEXTBOOK		Sections 12.0-12.6	Sections 12.7-12.11	Sections 12.12-12.15; chapter questions	Catch up/ study day
LAB MANUAL					
TEST BOOKLET	Test #4				

WEEK 13	DAY ONE	DAY TWO	DAY THREE	DAY FOUR	DAY FIVE
TEXTBOOK	Sections 13.0-13.4	Sections 13.5-13.7	Sections 13.8-13.11; chapter questions	Catch up/ study day	Sections 14.0-14.4
LAB MANUAL					
TEST BOOKLET					

WEEK 14	DAY ONE	DAY TWO	DAY THREE	DAY FOUR	DAY FIVE
TEXTBOOK	Sections 14.5-14.9	Sections 14.10-14.17	Sections 14.18-14.20; chapter questions		Sections 15.0-15.4
LAB MANUAL					
TEST BOOKLET				Test #5	

WEEK 15	DAY ONE	DAY TWO	DAY THREE	DAY FOUR	DAY FIVE
TEXTBOOK	Sections 15.5-15.9	Sections 15.10-15.13; chapter questions	Catch up/ study day	Sections 16.0-16.5	Sections 16.6-16.10
LAB MANUAL					
TEST BOOKLET					

WEEK 16	DAY ONE	DAY TWO	DAY THREE	DAY FOUR	DAY FIVE
TEXTBOOK	Sections 16.11-16.17; chapter questions	Catch up/ study day		Sections 17.0-17.6	Sections 17.7-17.9
LAB MANUAL					
TEST BOOKLET			Test #6		

WEEK 17	DAY ONE	DAY TWO	DAY THREE	DAY FOUR	DAY FIVE
TEXTBOOK	Sections 17.10-17.13	Sections 17.14-17.16; chapter questions	Catch up/ study day	Sections 18.0-18.4	Sections 18.5-18.10
LAB MANUAL					
TEST BOOKLET					

WEEK 18	DAY ONE	DAY TWO	DAY THREE	DAY FOUR	DAY FIVE
TEXTBOOK	Sections 18.11-18.17; chapter questions	Catch up/ study day	Sections 19.0-19.5	Sections 19.6-19.10	Sections 19.11-19.13; chapter questions
LAB MANUAL					
TEST BOOKLET					

WEEK 19	DAY ONE	DAY TWO	DAY THREE	DAY FOUR	DAY FIVE
TEXTBOOK	Catch up/ study day		Sections 20.0-20.6	Sections 20.7-20.10	Sections 20.11-20.14; chapter questions
LAB MANUAL					
TEST BOOKLET		Test #7			

WEEK 20	DAY ONE	DAY TWO	DAY THREE	DAY FOUR	DAY FIVE
TEXTBOOK	Catch up/ study day	Sections 21.0-21.7	Sections 21.8-21.11	Sections 21.12-21.14; chapter questions	Catch up/ study day
LAB MANUAL					
TEST BOOKLET					

WEEK 21	DAY ONE	DAY TWO	DAY THREE	DAY FOUR	DAY FIVE
TEXTBOOK		Sections 22.0-22.4	Sections 22.5-22.7	Sections 22.8-22.10	Sections 22.11-22.15; chapter questions
LAB MANUAL				Lab 5, or	Lab 5
TEST BOOKLET	Test #8				

WEEK 22	DAY ONE	DAY TWO	DAY THREE	DAY FOUR	DAY FIVE
TEXTBOOK	Catch up/ study day	Sections 23.0-23.5	Sections 23.6-23.10	Sections 23.11-23.15; chapter questions	Catch up/ study day
LAB MANUAL					
TEST BOOKLET					

WEEK 23	DAY ONE	DAY TWO	DAY THREE	DAY FOUR	DAY FIVE
TEXTBOOK		Sections 24.0-24.7	Sections 24.8-24.12	Sections 24.13-24.16	Sections 24.17-24.20
LAB MANUAL		Lab 6, or	Lab 6		Lab 7
TEST BOOKLET	Test #9				

WEEK 24	DAY ONE	DAY TWO	DAY THREE	DAY FOUR	DAY FIVE
TEXTBOOK	Catch up/ study day	Sections 24.21-24.23; chapter questions	Catch up/ study day	Sections 25.0-25.5	Sections 25.6-25.8
LAB MANUAL		Lab 8		Lab 9, or	Lab 9
TEST BOOKLET					

WEEK 25	DAY ONE	DAY TWO	DAY THREE	DAY FOUR	DAY FIVE
TEXTBOOK	Sections 25.9-25.11	Sections 25.12-25.16; chapter questions	Catch up/ study day		Sections 26.0-26.6
LAB MANUAL	Lab 10, or	Lab 10			
TEST BOOKLET				Test #10	

WEEK 26	DAY ONE	DAY TWO	DAY THREE	DAY FOUR	DAY FIVE
TEXTBOOK	Sections 26.7-26.12	Sections 26.13-26.17; chapter questions	Catch up/study day	Sections 27.0-27.5	Section 27.6
LAB MANUAL					Lab 11
TEST BOOKLET					

WEEK 27	DAY ONE	DAY TWO	DAY THREE	DAY FOUR	DAY FIVE
TEXTBOOK	Section 27.7	Catch up/study day	Sections 27.8-27.9	Sections 27.10-27.12; chapter questions	Catch up/study day
LAB MANUAL	Lab 12		Lab 13		
TEST BOOKLET					

WEEK 28	DAY ONE	DAY TWO	DAY THREE	DAY FOUR	DAY FIVE
TEXTBOOK		Sections 28.0-28.4	Sections 28.5-28.6	Sections 28.7-28.8	Sections 28.9-28.12; chapter questions
LAB MANUAL			Lab 14		
TEST BOOKLET	Test #11				

WEEK 29	DAY ONE	DAY TWO	DAY THREE	DAY FOUR	DAY FIVE
TEXTBOOK	Catch up/study day	Sections 29.0-29.4	Sections 29.5-29.7	Sections 29.8-29.10	Sections 29.11-29.14
LAB MANUAL				Lab 15	Lab16 Lab 17 (optional)
TEST BOOKLET					

WEEK 30	DAY ONE	DAY TWO	DAY THREE	DAY FOUR	DAY FIVE
TEXTBOOK	Sections 29.15-29.17; chapter questions	Catch up/study day		Sections 30.0-30.3	Sections 30.4-30.9
LAB MANUAL					
TEST BOOKLET			Test #12		

WEEK 31	DAY ONE	DAY TWO	DAY THREE	DAY FOUR	DAY FIVE
TEXTBOOK	Sections 30.10-30.12	Sections 30.13-30.16	Sections 30.17-30.19; chapter questions	Catch up/ study day	Sections 31.0-31.3
LAB MANUAL		Lab 18			
TEST BOOKLET					

WEEK 32	DAY ONE	DAY TWO	DAY THREE	DAY FOUR	DAY FIVE
TEXTBOOK	Sections 31.4-31.6	Sections 31.7-31.9	Sections 31.10-31.12; chapter questions	Catch up/ study day	
LAB MANUAL					
TEST BOOKLET					Test #13

WEEK 33	DAY ONE	DAY TWO	DAY THREE	DAY FOUR	DAY FIVE
TEXTBOOK	Sections 32.0-32.3	Sections 32.4-32.5	Sections 32.6-32.7	Sections 32.8-32.9; chapter questions	Catch up/ study day
LAB MANUAL					
TEST BOOKLET					

WEEK 34	DAY ONE	DAY TWO	DAY THREE	DAY FOUR	DAY FIVE
TEXTBOOK	Sections 33.0-33.4	Sections 33.5-33.7	Sections 33.8-33.10	Sections 33.11-33.13; chapter questions	Catch up/ study day
LAB MANUAL			Lab 19		
TEST BOOKLET					

WEEK 35	DAY ONE	DAY TWO	DAY THREE	DAY FOUR	DAY FIVE
TEXTBOOK		Sections 34.0-34.6	Sections 34.7-34.10	Sections 34.11-34.13	Sections 34.14-34.17
LAB MANUAL			Lab 20		
TEST BOOKLET	Test #14				

WEEK 36	DAY ONE	DAY TWO	DAY THREE	DAY FOUR	DAY FIVE
TEXTBOOK	Sections 34.18-34.19	Sections 34.20-34.22; chapter questions	Catch up/ study day		
LAB MANUAL					
TEST BOOKLET				Test #15	

Study Question Answer Key

CHAPTER 1

1. All living organisms: Contain DNA; are made of one or more cells; reproduce; are complex and organized; are responsive to their environment; extract energy from their surroundings; maintain homeostasis; grow; are scientifically classified.

2. A food chain is the process or relationship, of the transfer of energy from the sun, to the producers, to the consumers, to the decomposers.

3. Plants obtain their sugar molecules by making them.

4. Herbivores obtain their daily energy by eating plants, or producers. Carnivores obtain their daily energy by eating meat.

5. Carnivores obtain their daily energy by eating only consumers, omnivores obtain their daily energy by eating producers or consumers.

6. The process of converting sugar molecules into energy molecules is called cellular respiration.

7. Homeostasis is the property of maintaining a stable internal environment. If homeostasis is not maintained, an organism cannot perform the proper chemical reactions, and it will die.

8. The producer, consumer, and decompser system; the herbivore, carnivore, and omnivore system; the seven level, six kingdom, taxonomic system that was described by Linnaeus.

9 The current system of classification is a seven level classification system that names the organism based on its kingdom, phylum, class, order, family, genus and species.

10. A hypothesis statement is an educated statement or guess that explains observations regarding a particular subject. A hypothesis statement is made early on in the scientific method.

11. After a hypothesis statement is made, data is gathered that specifically addresses the hypothesis.

12. In a controlled experiment, the experimental group is exposed to the variable.

13. If the data does not support the hypothesis, the hypothesis either needs to be restated in a way that is consistant with the data, or thrown out altogether.

14. SI units are important because all scientists understand them.

15. Magnification is the amount or degree which an image is enlarged by a microscope. Resolution is the ability to see an image's fine detail.

16. Scanning electron microscopes provide three-dimensional images of objects.

CHAPTER 2

1. Sugars are the universal fuel source.

2. Plants make sugar (glucose) by capturing and using the sun's energy.

3. The process in question 2 is called photosynthesis.

4. Cellular respiration is the set of biochemical reactions that consumers use to make their energy molecules.

5. Matter is anything that takes up space and has mass. Mass is a measurement of matter, whereas matter is the substance you are discussing.

6. The three subatomic particles are protons, neutrons, and electrons.

7. An element is matter composed only of one type of atom. More then 90% of all life mass is composed of carbon, oxygen, hydrogen, and nitrogen.

8.

9. The majority of an atom's mass is contained in the nucleus.

10. False.

11.

Symbol	Atomic #	Protons	Neutrons	Electrons
H	1	1	0	1
Ne	10	10	10	10
Ni	28	28	30	28
Bi	83	83	126	83
K	19	19	20	19
Gd	64	64	93	64
Mg	12	12	12	12

12. The number of protons in a neutral atom always equals the number of electrons.

13. False.

14. Hydrogen 1, Carbon 2, Sodium 3, Silicone 3, Chlorine 3, Oxygen 2.

15. One answer could be that electrons prefer to orbit the nucleus in pairs. Another answer could be that electrons prefer to orbit the nucleus in shells or clouds.

16. An isotope is a different form of the same element, differing only in the number of neutrons each form contains.

17. Atoms prefer to bond with other atoms, such that each one contains eight electrons in its outermost shell. This is referred to as the octet rule.

18. A covalent chemical bond is a bond that forms between two atoms in which the electrons involved in the bond are equally shared between the two atoms. An ionic bond forms between a positive ion and a negative ion.

19. $C + O_2 \longrightarrow CO_2$

20. C and O_2 are the reactants and CO_2 is the product.

21. A charged atom is called an ion.

22. True.

23. When sodium chloride (NaCl) is mixed in water it causes it to dissociate into the Na^+ and Cl^- ions.

$$NaCl \xrightarrow[H_2O]{} Na^+ + Cl^-$$

24. The three important properties we discussed about water, are that water: Is a polar molecule, or forms through polar covalent bonds; Is a non-linear molecule; forms hydrogen bonds.

25. In a solution, the part of the solution that is greater in concentration is called the solvent. The part of the solution that is in lesser concentration is called the solute.

26. A solution with a low pH is an acidic solution. It contains more hydrogen ions than hydroxide ions.

CHAPTER 3

1. Organic compounds are made by living organisms and are molecules that utilize carbon as a backbone and also commonly contain hydrogen and oxygen.

2. "Carbon compound" may also be used since all organic molecules contain carbon and are built upon a carbon backbone. In addition, organic molecules are not only synthesized in living organisms; they can now be made in the laboratory.

3. Organic molecules: Are synthesized by all living things. Contain carbon and hydrogen. Most (not all) contain oxygen. Are built from monomers.

4. True.

5. False. This chemical formula describes many molecule types due to isomers.

6. Writing the structural formula for a molecule resolves the problems created by isomerization.

7. C -C is a covalent bond. C=C is a double covalent bond.

8. C_2H_2 H —— C≡≡C —— H

9. Answers may vary as there are many, many combinations. Several are shown below.

10. Larger polymers are synthesized from smaller monomer subunits via condensation (or dehydration) synthesis reactions.

   or

   Larger organic molecules are synthesized from smaller monomer subunits via condensation (or dehydration) synthesis reactions.

11. Water is always a product of a condensation (dehydration) synthesis reaction.

12. Water is added across the monomer/polymer bond, which releases the monomer unit from the polymer.

13. CH_2 refers to the general chemical formula for a carbohydrate molecule.

14. A glycoside bond.

15. Glycogen is a carbohydrate macromolecule. It is a polymer of glucose and is the form of energy storage that animals use.

16. True.

17. True.

18. $COOH^-$

19. Fatty acids are made up of a glycerol backbone with fatty acids attached to it.

20. The monomeric unit of a protein is an amino acid.

21. This is a condensation (dehydration) synthesis reaction. The hydroxide group from the carboxylic acid of one amino acid monomer and a hydrogen atom from the nitrogen (amino) group of the other amino acid monomer are removed. This results in a peptide bond between the two amino acids and the formation of water.

22. Nucleic acids are larger than nucleotides.

23. A phosphate group, central sugar and a nitrogen containing (nitrogenous) base.

CHAPTER 4

1. Cell theory attempts to explain what has been observed about cells. It states that cells are the basic units of structure and function of all life, that all living things are made up of one or more cells and that all cells are produced from other cells.

2. "Round" and "square" are two-dimensional descriptors but cells are three-dimensional objects. Sphere is a three-dimensional word that better decribes the shape of a "round" cell and cube is a three-dimensional descriptor of a "square" cell.

3. False. There are many unicellular life forms.

4. Trillions (at least one trillion).

5. The functions of the cell are the same as the properties of life. All cells: contain DNA; reproduce; are complex and organized; respond to the environment; extract energy from their surroundings; maintain homeostasis; grow; absorb, digest, and process nutrients; biosynthesize; perform transport; protect from harm.

6. False. Not all cells have a cell wall.

7. Prokaryotic.

8. The integral and peripheral membrane proteins.

9. Cell walls serve as an added layer of protection and support.

10. Compare the student's drawing to Figure 4.22.1

11. The structure of the phospholipid molecules with the hydrophilic heads and hydrophobic tails.

12. False.

13. The integral protein needs to contain a hydrophilic area and a hydrophobic area.

14. A hydrophilic molecule would congregate at the surface and a hydrophobic molecule in the middle of the lipid bilayer.

15. Diffusion, osmosis, and gated ion channel transport are types of passive transport. Gated channel transport requires integral proteins to perform.

16. Both involve the random movement of molecules down their concentration gradients across a semi-permeable membrane. They are different because osmosis specifically involves the movement of a solvent from an area of low solute concentration, to an area of high solute concentration.

17. The distinguishing feature is whether or not energy needs to be used to perform the transport. Passive transport requires no energy, active transport does.

18. In an isotonic environment, the muscle cell would maintain its normal shape and function. In a hypertonic environment, the cell would lose its water and would shrivel. In a hypotonic environment, the cell would take up water, swell, and possibly burst.

19. They are both forms of endocytosis (so they both involve vesicle formation). They are different because phagocytosis brings solid particles and molecules into the cell and pinocytosis brings liquid into the cell.

20. A vesicle.

CHAPTER 5

1. Eukaryotic cells have more kinds of organelles than prokaryotes; eukaryotes contain membrane-bound organelles, prokaryotes do not; Eukaryotic DNA is in the nucleus; Prokaryotic DNA is free in the cytoplasm. All multicellular organisms are eukaryotic. All prokaryotic organisms are unicellular.

2. An organelle is the individual functional unit of the cell.

3. Protoplasm is composed of cytoplasm and the nucleus, or, protoplasm is composed of cytoplasm, organelles and the nucleus.

4. False. The cytosol is 70% water and 30% ions, proteins and other molecules.

5. Nucleoplasm and cytoplasm.

6. The cytoskeleton is a complex meshwork of proteins running throughout the cytoplasm; The cytoskeleton holds the cell together; The cytoskeleton serves as a highway for the transport of substances; The cytoskeleton is composed of microtubules (tubulin), intermediate fibers, and microfilaments (actin).

7. Centrioles, eukaryotic flagella, and cilia.

8. DNA and the nucleolus; The nucleolus manufactures ribosomal parts/components.

9. Substances move in and out of the nucleus through nuclear pores.

10. From ribosome, to endoplasmic reticulum, to golgi apparatus, to the cell.

11. The double membrane-bound organelles are the nucleus, mitochondrion, and plastid (chloroplast is an acceptable answer for plastid).

12. A storage plastid; You are standing in a plant cell because animal cells do not contain plastids.

13. Lysosomes and peroxisomes both: Are membrane-bound organelles; Contain toxic substances; Function to keep the cytoplasm clean.

14. The middle lamella is the sticky substance secreted by plant cells that hold all plant cells together.

15. Animal cells are held together by the extracellular matrix (ECM).

16. If you are in a cell and you see a nucleoid, you are in a prokaryotic cell because the nucleoid is where prokaryotic DNA is kept.

CHAPTER 6

1. Energy is the ability to do (or perform) work.

2. In general, the summation of an organism's anabolic and catabolic processes is called metabolism.

3. Potential energy is the ability to perform work, "stored" energy. Kinetic energy is the energy of motion, or "active" energy.

4. Mechanical and chemical energy are types of energy. Most of the energy in living organisms exists as chemical energy.

5. The study of energy transfer is called thermodynamics.

6. The First Law of Thermodynamics states energy can neither be created nor destroyed. Chemical reactions of life result in the transfer of one type of energy to another. During these reactions, some of the energy is transferred in the form of potential energy in the bonds of the newly formed molecule(s) and the rest is lost as heat .

7. The Second Law of Thermodynamics states that chemical reactions and systems favor increasing entropy. An individual human being is considered a system. Humans (and all organisms) need to perform additional chemical reactions to counteract the increasing entropy to maintain homeostasis (in other words to maintain order in the system). Further, human entropy is in the form of heat, called "body heat." This "body heat", allows humans to maintain temperature homeostasis.

8. Endergonic chemical reactions are used because they are able to be tightly controlled due to the need to put energy into the reaction to get it started.

9. The first reaction is an exergonic reaction since it occurred spontaneously and did not require an input of energy. In the second example, heating the water is the equivalent of providing the activation energy for the reaction to occur. This an example of an endergonic reaction.

10. An enzyme lowers the activation energy of the reaction the enzyme catalyzes.

11. The active site is that part of the enzyme that binds directly to the substrate. The active site creates a microenvironment that makes the conditions of the reaction more favorable to occur.

12. The lock and key hypothesis does not adequately explain the structural change that occurs in an enzyme when it binds the substrate. The induced fit model takes into account both the lock and key fit of an enzyme and substrate and also the structural change the enzyme undergoes after binding the substrate.

13. True.

14. Three.

15. False. ATP provides all of the energy a cell needs to perform work.

16. Enzymes break the phosphate bonds of ATP, which releases energy. The enzyme is able to capture the released energy and couple it to the reaction the enzyme catalyzes. In this way, the enzyme harnesses the released energy and uses it to provide the activation energy to start the reaction.

17. Enzymes know when to turn on and off through the process of feedback inhibition.

CHAPTER 7

1. Carbon dioxide is removed from and oxygen released into the air, and carbon fixation occurs.

2. Carbon fixation is the incorporation of carbon atoms into organic molecules.

3. The biological term for producer is autotroph, of which there are two types - photoautotrophs and chemoautotrophs. Photoautotrophs produce their cell mass and organic molecules by converting sunlight energy and chemoautotrophs do so by converting chemical energy obtained from inorganic molecules.

4. Radiowaves have the longest wavelength, then microwaves, then infrared, then visible light, then ultraviolet, xrays, and, finally, gamma rays have the shortest wavelengths.

5. A photon is the individual unit of electromagnetic energy. They are packets of electromagnetic energy and travel in waves.

6. False. The difference in wavelengths is what serves as the differentiating factor between the seven types of electromagnetic waves.

7. False.

8. The visible spectrum includes all wavelengths of electromagnetic radiation that can be seen by the naked eye. It includes the wavelengths 380 nm through 750 nm.

9. False. When light is passed through plant leaves, the only colors that emerge are green and yellow, which means they were not absorbed.

10. False. Energy that is reflected cannot be used to drive photosynthesis. Only absorbed energy can be retained and converted by the plant to drive the chemical reactions of photosynthesis.

11. Compare the drawing to Figure 7.10.1. Photosynthesis occurs in the chloroplasts.

12. Six molecules of water plus six molecules of carbon dioxide plus (or using) light energy from the sun results in the formation of one molecule of glucose and six molecules of oxygen.

13. The two groups of reactions are the light reactions (or light-dependent reactions) and the Calvin cycle (or light-independent reactions). The products of the light reactions are ATP and NADPH. The product of the Calvin cycle is G-3-P (or glucose is an acceptable answer).

14. They are both photosynthetic pigments. However, they absorb light of different wavelengths.

15. It is a series (or group) of proteins and enzymes that pass high energy electrons between one another. The electrons move from higher to lower energy states. At key points in the chain, the energy being lost by the electron is captured and used to make some useful molecule.

16. The Calvin cycle generates molecules of G-3-P (which is then usually converted into glucose). The energy for the Calvin cycle comes from the light-dependent reactions; specifically, the electron transport chain between PS II and PS I (which produces ATP) and the chain after PS I (which produces NADPH).

CHAPTER 8

1. Cellular respiration is the set of biochemical reactions that organisms use to break down (or metabolize) glucose and harness the released energy to form ATP. Cellular respiration is the main way ATP is formed, which is then used to provide the fuel for the endergonic reactions of life.

2. The phosphate bonds of ATP store energy. When the phosphate bonds are broken, enzymes harness the released energy and use it to fuel the endergonic reaction the enzyme catalyzes.

3. ATP provides the energy needed to start an endergonic reaction. That means ATP is the source of the activation energy for endergonic reactions.

4. When the first phosphate bond of ATP is broken, energy is released, an inorganic phosphate group is removed, and ADP is formed.

5.

$$ATP + H_2O \xrightarrow{\text{Enzyme}} ADP + Pi + Energy$$

6.

$$C_6H_{12}O_6 + 6 O_2 \longrightarrow 6 CO_2 + 6 H_2O + 36 ATP$$

7. Aerobic cellular respiration.

8. Anaerobic cellular respiration, or fermentation.

9. Aerobic cellular respiration is much more efficient than anaerobic respiration. That means that many more ATP molecules are made per glucose molecule burned during aerobic cellular respiration than during anaerobic.

10. Glycolysis is the biochemical process that converts glucose into two molecules of pyruvate. The products are ATP, NADH, and pyruvate.

    It is important as the first step of aerobic and anaerobic respiration.

11. False.

12. False. Glycolysis is not technically part of aerobic respiration because glycolysis occurs at other times than during aerobic respiration. Although it is true that aerobic respiration cannot occur without gycolysis first occurring, glycolysis is not considered to be part of aerobic cellular respiration.

13. False. Glycolysis can occur in the presence or absence of oxygen.

14. True.

15. See Figure 8.7.1 for the drawing and labels.

    Cristae are formed from the inner membrane and contain the electron transport chain and produce the majority of ATP. Outer membrane is for protection. The matrix is where the transition reaction and Krebs cycle occur.

16. During an oxidation reaction, a hydrogen ion with its electon is removed from a molecule.

17. During pyruvic acid oxidation, one of the carbon atoms is removed from the pyruvate, forming carbon dioxide. The other two carbons are attached to a molecule of co-enzyme A (co-A) forming a molecule of acetyl co-enzyme-A (acetyl co-A). The hydrogen ion with its electron removed from pyruvate is added to a molecule of NAD, forming NADH.

18. Acetyl co-enzyme A carries the two remaining carbon atoms from the pyruvate molecule through the Krebs cycle. 2 molecules of carbon dioxide are formed, 1 molecule of ATP, 3 molecules of NADH and 1 molecule of $FADH_2$ are formed for each turn of the cycle. The Krebs cycle occurs twice for each molecule of glucose that enters cellular respiration.

19. NAD and FAD.

20. The electron transport chain produces far more ATP molecules than the Krebs cycle. 32 ATP molecules are formed in the electron transport chain versus the 2 formed in the Krebs cycle.

21. Fermentation.

22. Anaerobic respiration can produce ATP in the absence of oxygen and it produces ATP more quickly than aerobic respiration can.

23. The proton pump is an intergral protein that pumps hydrogen ions against their gradient from one side of the cristae to the other. This causes a high proton concentration on one side of the membrane. This concentration gradient has the ability to perform work, which is called the proton motive force.

    Chemiosmosis is the process of hydrogen ions moving back across the cristae membrane, providing the energy for ATP synthase to convert to ATP. When the protons move through the ATP synthase, they generate energy, which is used by ATP synthase to catalyze the endergonic makes ATP from ADP.

CHAPTER 9

1. DNA is deoxyribonucleic acid and RNA is ribonucleic acid.

2. The four nucleotides of DNA are adenine, cytosine, thymine and guanine. Adenine and thymine base pair with one another while cytosine and guanine base pair with one another.

3. DNA is a double-stranded helix.

4. There are four nucleotides that form the structure of DNA. They are linked together repeatedly to form the DNA macromolecule.

5. False. Adenine (and guanine) are purines. Thymine (and cytosine) are pyrimidines.

6. True.

7. The genetic code is the specific linear sequence of nucleotides of DNA within a gene segment. The individual unit of the genetic code is the codon. Each codon codes for the production (or insertion) of one specific amino acid into a protein. Therefore the genetic code contains the information for the production of proteins.

8. TAC GAA TTG GCT

9. A chromosome is a long segment of DNA which is wrapped around histones and further condensed and coiled so that it is a very dense piece of DNA.

10. Chromosomes and chromatin are essentially the same thing. The difference is in what the physical structure of the DNA looks like. Chromatin is essentially a chromsome that is very loosely coiled. An actual "chromosome" is the form of DNA that is visible as a tightly coiled molecule.

11. False. Histones are the special proteins that DNA is wrapped around so that it can be tightly coiled and take up less space in the nucleus.

12. Chromosomes and genes are both composed of DNA. DNA is housed in smaller pieces called chromosomes. Genes are smaller segments within a chromosome. Each gene codes for the production of a protein.

13. DNA contains the central sugar deoxyribose, RNA contains ribose. DNA is a double-stranded molecule, RNA is single-stranded. DNA contains the base thymine, RNA contains uracil.

14. - DNA is unzipped by helicase
    - RNA polymerase binds to DNA and makes mRNA from the DNA template
    - mRNA falls off the DNA when completed
    - mRNA introns are removed and exons are spliced together
    - mRNA moves from the nucleus to the cytoplasm
    - ribosomes bind to the mRNA in the cytoplasm
    - ribosomes read mRNA codons in a linear sequence and base pair mRNA codons to tRNA anticodons
    - ribosomes catalyze the dehydration synthesis reactions that link amino acids together
    - ribosomes move along the mRNA in a linear sequence continually base pairing codons and anticodons, adding the appropriate amino acids to the growing protein until the mRNA is completely translated

15. tRNA looks like a "T" or cross because of hydrogen bonding that occurs between the nucleotides of the RNA molecule.

16. The mRNA codon and tRNA anticodon must be complementary for the ribosome to know that the tRNA is carrying the correct amino acid. If the mRNA codon and the tRNA anticodon are complementary, then the ribosome is able to base pair the two together. This tells the ribosome that the tRNA is carrying the correct amino acid.

17. Phe-Thr-Ser-Ala-Trp

18. Asp-Gly-Asn-Tyr-stop

CHAPTER 10

1. Cell division is the biological process of one cell dividing into two cells. The two types of cell division are sexual and asexual cell division (reproduction).

2. A parent cell is the "starting" cell in cell reproduction. It first replicates its DNA and then divides into two daughter cells.

3. False. Normally, after cell division is completed, both daughter cells have a fully functional copy of DNA.

4. The purpose of mitosis is to form two fully functional cells, each with a fully functional copy of DNA, from one single cell (or parent cell).

5. Spontaneous generation is the idea that non-living substances can form or produce living organisms. Spontaneous generation does not occur.

6. The Theory of Biogenesis states that only living organisms can produce living organisms. Therefore, all living organisms are produced by living organisms.

7. False. The enzyme helicase breaks the hydrogen bonds between the two strands of DNA. This unzips and unwinds the DNA.

8. DNA polymerase forms a new daughter strand across from a parent strand. It base pairs and links together nucleotides of the daughter strand using the nucleotide sequence of the parent strand as a template.

9. See Figure 10.7.3. The base pairs on the parent DNA strand and on the daughter strand must be complementary (T across from A and G across from C). In Figure 10.7.3, the black strands are the parent strands and the red are the daughter strands.

10. No. Mitosis is a form of asexual reproduction.

11. Helicase binds to the replication origin and begins to unzip/unwind DNA at that location.

12. Interphase and mitosis are the two phases of the cell cycle.

13. The phases of mitosis are prophase, prometaphase, metaphase, anaphase, and telophase.

    Prophase begins after the chromosomes are copied in interphase. The nucleoli disappear and the DNA condenses into formal chromosomes. Centrioles form and begin to assemble the spindle.

    In prometaphase, the nuclear membrane disintegrates, the spindle fully forms and attaches to the centromeres of each pair of sister chromatids. The fully condensed chromosomes begin to align along the equator of the cell.

    During metaphase, the chromosomes align completely along the equator of the cell.

    In anaphase, the spindle fibers contract, which results in karyokinesis (pulling the sister chromatids apart).

    In telophase, cytokinesis occurs, the chromosomes begin to unwind, and the nuclear membrane reforms around the DNA (chromosomes).

14. False. The cell spends 90% of its time in interphase.

15. The spindle has two functions. The spindle moves the chromosomes to align along the equator of the cell and then when it contracts completely, it pulls the chromosomes to opposite ends of the cell.

16. False. Karyokinesis is the name given to the process of pulling chromosomes apart during cell division. Cytokinesis is the name given to the process of the cell membrane pinching inward and forming two cells from one cell.

17. A cell plate is a structure that forms between two plant cells after the parent cell divides into two daughter cells during mitosis. The cell plate later becomes the cell wall between the two new daughter cells.

18. Cancer cells do not demonstrate the property of stopping cell division when they come into contact with one another. In addition, many cancer cell lines have lost the property of apoptosis.

19. During interphase, the level of cyclin slowly builds. It eventually reaches a certain critical level, at which time it then binds to a type of enzyme called a kinase. This specific kinase is called cyclin-dependent protein kinase, or CDPK. CDPK becomes active when cyclin binds to it. The activated CDPK then performs a chemical reaction called a phosphorylation reaction and adds a phosphate group to a particular protein. The level of this protein then increases. When this protein reaches a certain critical level, it stimulates the cell to enter mitosis.

CHAPTER 11

1. False. Almost all bacterial cells have only one chromosome.

2. True.

3. False. All organisms of the same species have the same number of chromosomes.

4. True.

5. Autosomes are chromosomes that contain the information (or code for) the development of body characteristics.

6. A karyotype is the orderly presentation of all of the chromosomes obtained from one cell. It is standard procedure to arrange the chromosomes in pairs in a karyotype.

7. Humans normally have one pair (or two individual) sex chromosomes. They are called the X and Y chromosomes.

8. True.

9. Sexual reproduction is the formation of a new organism through the combination of genetic material (or DNA or chromosomes) from the male and female parent.

10. An n cell is also called a haploid cell. It contains a complete set of unpaired chromosomes. A cell which is 2n is also called a diploid cell. It contains a complete set of paired chromsomes (one set from the male parent and the other set from the female parent).

11. The n number for the fruit fly is 4, and the 2n is 8. For the sand dollar, n is 26 and 2n is 52.

12. True.

13. Meiosis is the process of a 2n, or diploid, reproductive cell undergoing cell division in such a way that four haploid gametes are formed. The purpose of meiosis is to ensure that each gamete receives the haploid number of chromosomes.

14. After meiosis I is completed, two cells have been formed. Each cell contains two copies of the n number of chromosomes (or each cell contains two copies of the haploid number).

15. Following meiosis II, four cells have been formed. These cells are called gametes and each one contains the haploid, or n, number of chromosomes.

16. Four.

17. One.

18. Polar bodies are three degenerative cells that form as a result of human female meiosis.

19. When the homologous pairs align during meiosis I (or when synapsis occurs), similar areas of DNA are very close to one another. As a result, small (or sometimes large) pieces of DNA can be exchanged between one chromosome and another. This results in the exchange of genetic information.

20. A clone is a cell genetically identical to another cell. Clones are formed as a result of asexual reproduction.

CHAPTER 12

1. A gene is the functional unit of heredity. One gene codes for the production of one protein. One gene controls one trait. It is important to understand the concept of genes and how they are passed from parent to offspring in order to understand hereditary patterns.

2. When a gene is expressed, the protein the gene codes for is being made by the cell/organism. Another way of stating this is that when a gene is expressed, the trait the gene codes for (or controls) is visible in some way.

3. If a gene codes for blue eyes, then when that gene is expressed, a person's eyes will appear blue.

4. Genome is the term used to mean the total number, or total amount, of genes an organism contains.

5. Genes are the individual units of heredity. Heredity occurs because genes are passed from one organism to another (from parent to offspring) as a result of sexual reproduction.

6. True.

7. A characteristic is a describable feature of an organism. A trait is the variant of the characteristic. For example, if you were studying fur color in bears, the characteristic you would be looking at is fur color. If there are two different colors of fur - black and brown - there would two variants of fur color. The characteristic is fur color, the variant is the two different colors, black and brown.

8. A trait is an identifiable characteristic of an organism (or, a trait is a variant of a characteristic). Traits are controlled by genes.

9. Gregor Mendel was the first person to study and learn how to predict the passage of traits from one generation to the next. He discovered and clearly identified basic genetic principles that are still in use today.

10. True.

11. The only way that a recessive allele can be expressed (or for a recessive trait to be expressed) is for the organism to be homozygous recessive for that particular condition. Another way of stating that is the organism must contain two recessive alleles for the same condition in order for the recessive condition to be expressed.

12. False. If an organism has one dominant allele and one recessive allele, the trait of the recessive allele is suppressed by the dominant allele. The trait of the dominant allele is expressed.

13. The genotype is the actual genetic makeup of an organism. The genotype controls the phenotype.

14. The P generation is the starting generation in a genetic study. Depending on the study, it can refer to the parental or pure generation. The F1 generation is the generation of organisms that are produced when the P generation organisms are bred with one another. The F2 generation is the generation of organisms that are produced when the F1 generation organisms are bred with one another.

15. Homozygous recessive is written "nn." The nail color is black.

16.

	n	n
N	Nn Brown	Nn Brown
N	Nn Brown	Nn Brown

17.

	N	n
N	NN Brown	Nn Brown
n	Nn Brown	nn Black

18. The specific letter the student chooses to designate the allele is not important (it could be "A" and "a," or "H" and "h," or "R" and "r"). What is important is that the dominant allele must be assigned the capital letter and recessive allele the lower case letter. For example, the "R" allele would be the dominant round shape/trait and the "r" allele would be the recessive oval shape/trait.

19. All of the offspring would have leaves that are round.

20. Independent assortment is the genetic property that genes and their respective chromosomes assort (or are distributed) independent of one another during meiosis. This occurs as a totally random process.

21. Gene segregation and independent assortment are Mendelian Principles.

CHAPTER 13

1. 50% chance to receive the dominant or recessive allele.

2. This is really a question of probability and requires you to use knowledge concerning sex-linked traits and diseases. Recall a sex-linked disease is caused by a defective allele carried on a sex chromosome. However, the allele codes for a non-sexual trait. The X chromosome is larger than the Y and so carries many more sex-linked traits than the Y chromosome can. In order for a boy to be born with a sex-linked disease, he only needs to inherit ONE defective X chromosome because there is no corresponding normal allele on the Y chromosome to counteract the abnormal allele on the X.

   In order for a girl to be born with a sex-linked disease, she would need to inherit two defective X chromosomes. Further, each defective X chromosome would need to contain the same defective allele. Therefore, it is far more likely that a family would have a boy born with a sex-linked disease rather than a girl born with one.

   In this particular case, we are told that a girl was born with this genetic disease. Since we know it is very unlikely for a girl to be born with a sex-linked disease, we can therefore predict that this particular disease is much more likely to be one carried on an autosome rather than a sex-linked disease.

3. The probability of having three children in a row with the genetic disease is 1 in 64 (or 1/64 or 1.6%). The probability of the next child having the disease is 1 in 4 (or ¼ or 25%).

4. Incomplete dominance is an inheritance pattern in which no forms of the gene are dominant over the others and the resulting phenotype is a blend of the individual allele phenotypes. In this situation, when a black dog and a white dog have puppies, we would expect the fur to be a blend of the coat colors of the two parents, or some shade of gray.

5. Incomplete dominance is an inheritance pattern in which no forms of the gene are dominant over the others and the resulting phenotype is a blend of the individual allele phenotypes. Like incomplete dominance, in codominance, no alleles are dominant over the others. However, instead of the resulting phenotype being a blend of the traits of the two individual phenotypes, the phenotype of each allele is expressed.

6. True.

7. Epistasis is a condition in which one or more genes that do not code for a trait affect the expression of the trait. These genes are called modifier genes. Epistasis occurs as a result of modifier genes.

CHAPTER 14

1. The two general ways are mutation and independent assortment.

2. False. RNA genes code for the production of tRNA and ribosomal RNA (rRNA). mRNA is made as a result of a gene that codes for the production of a protein being transcribed.

3. False. Less than 1% of the total amount of human DNA actually codes for protein production. The remaining DNA codes for the production of RNA or has no known function.

4. Tandem repeats are DNA segments containing two or more nucleotides that are repeated over and over again along a continuous segment of DNA. Tandem repeats do not code for the production of proteins; their function is unknown.

5. A gene mutation is a change in the normal sequence of a gene. The large majority of DNA that does not code for protein production protects against mutation because when a mutation occurs, it is far more likely to occur in a segment of DNA that does not code for the production of a protein than to occur within a gene.

6. Many mutations occur within the large portion of DNA that does not code for protein production. Also, silent mutations do not result in alteration of the protein or the function of the protein.

7. Addition mutation.
   Deletion mutation.
   Substitution mutation.

8. True.

9. If a point mutation causes an amino acid codon to be mutated into a stop codon, it is called a nonsense mutation.

10. A missense mutation is a type of substitution mutation that causes an incorrect amino acid to be inserted into a protein.

11. True.

12. False. The type of mutation being described is a germ cell mutation. Organisms formed from germ cell mutations carry the mutation in every cell of the body.

13. 50% chance.

14. There are several answers for this question depending upon your viewpoint. Normal genetic variation is produced as a result of independent assortment and crossing over. Abnormal variation is produced as a result of mutation. Normal genetic variation results in normal variation of the phenotype. Abnormal genetic variation results in abnormal variation of the phenotype. Normal genetic variation is good for individuals and populations but abnormal genetic variation is not.

15. False. Germ cell mutations may be passed to offspring.

16. A nondisjunction chromosomal mutation occurs during meiosis I when the homologous pairs are aligned across from one another. During anaphase I, the chromosomes do not separate properly. Instead of one pair separating from the other and moving to opposite ends of the cell, usually one chromosome goes to one end of the cell and three chromosomes go to the other end of the cell.

CHAPTER 15

1. A geneticist is a scientist who studies genetics.

2. A genetic disease is a disease that is caused by a germ cell mutation which results in either a chromosomal mutation, or a gene mutation.

3. Once the normal human genome sequence is known, then it will be very easy to understand and identify what an abnormal human genome looks like. This will make it very easy to identify abnormal genes and how they may relate to a variety of human diseases.

4. False. Most genetic diseases are caused by recessive alleles.

5. This person is called a carrier. This person would not have the disease because the trait for the normal copy of the gene dominates the trait for the recessive copy of the gene.

6. Pedigree analyses are helpful in understanding genetic diseases because they allow one to track one trait or several traits very easily from generation to generation. This is the key feature of understanding how a genetic disease is transmitted in families.

7. False. Cystic fibrosis is a genetic disease in which 1:25 people of European Caucasian descent are affected.

8. 100% of the offspring will have cystic fibrosis. (So 100% chance all children will have cystic fibrosis.)

9. CF usually results from a point mutation in the gene which codes for a transmembrane protein which is responsible for transporting chloride across the membrane.

10.

	s	s
S	Ss	Ss
S	Ss	Ss

11. Some recessive alleles exist because people who are carriers of those alleles are protected against certain disease. This is called heterozygous superiority.

12. Children that are born to parents who have a genetic disease that is transmitted in a dominant fashion have a higher likelihood of having the disease because they only need to inherit that one dominant allele in order to have the disease.

13. False. Sex-linked genetic diseases always affect males more than females.

14. True.

15. Chorionic villus sampling, amniocentesis, ultrasound and fetoscopy are all ways of evaluating a baby for congenital birth defects before they are born.

CHAPTER 16

1. The many functions of DNA technology include: introducing a gene into an organism in order to give the organism a trait it did not previously have; RFLP analysis/DNA fingerprinting; polymerase chain reaction (PCR); gene isolation; selective breeding; gene cloning; organism cloning; gene therapy; and forensics.

2. Recombinant DNA is DNA that is derived from two or more organisms.

3. Reverse transcriptase is unusual because it has the ability to build (or synthesize) a molecule of DNA using an mRNA template. This is the enzyme that allows geneticists to work backwards. Working backwards is very important in recombinant technology.

4. Restriction enzymes cut DNA at very precise nucleotide sequences and when they cut DNA, sticky ends are created.

5. Vectors are required to deliver, or introduce, the recombinant DNA into the host. The two most commonly used vectors are plasmids and viruses.

6. True.

7. First, the gene of interest is isolated. Then the DNA containing the gene is cut by the appropriate restriction enzyme. The vector DNA is also cut with the same restriction enzyme. This creates complementary sticky ends on the vector and gene DNA. The two types of DNA are mixed together with the enzyme ligase. The ligase links the sticky ends together. This process forms a circular piece of recombinant DNA and is repeated thousands to millions of times, or more. The recombinant DNA is then packaged into the appropriate vector and introduced into the host.

8. In order to perform DNA fingerprinting, a process called restriction fragment length polymorphism is used. The first step is to cut a larger piece of DNA into smaller pieces with restriction enzymes. The smaller pieces are called restriction fragment lengths. The restriction fragments are poured onto a gel and subjected to an electrical current in a process called gel electrophoresis. During electrophoresis, the fragments of differing lengths and charges are driven different distances into the gel. After that, a radioactive DNA probe is added to the gel. This probe sticks to the DNA fragments. This causes the area of the gel where the restriction fragments are located to light up in bands. The banding patterns are referred to as the DNA fingerprint.

9. PCR refers to the process called polymerase chain reaction. PCR is used to amplify small amounts of DNA into large amounts. In PCR, the sample DNA is mixed with A,T,C, and G, as well as DNA polymerase and primers. After these materials are mixed together, the DNA polymerase begins to make new copies of DNA using the sample copies as templates. This results in the synthesis of many thousands (or more) molecules of DNA from the original sample.

10. Since sickle cell anemia and cystic fibrosis are caused by single gene defects, it is theoretically possible to cure these diseases using DNA technology. This would entail introducing the normal gene into the cells of the person with the genetic disease. Once the normal gene is inside the cells of the person with the genetic disease, they could start to make the normal gene product and their genetic disease would be cured.

CHAPTER 17

1. The two main theories are creation and evolution.

2. Creationists believe all life was created by God, upon His word, for a specific reason. Evolutionists believe that all life on earth arose as a a result of a random process in which less complex organisms directly gave rise to more complex organisms over billions of years.

3. An evolutionist is a person who believes that evolution accurately explains the origin and diversity of life on earth. A creationist is a person who believes that creation accurately explains the origin and diversity of life.

4. False. Your interpretation of data critically depends upon your philosophy regarding the nature of that data.

5. Four fossil types are casts, molds, amber, and tracks. A cast is formed when an organism is covered by sediments. The parts of the organism break down and are replaced by minerals, which solidify and take the shape and appearance of the original organism. A mold forms after an imprint of the organism is made on the sediment. Amber is tree sap that traps organisms and preserves them. Tracks are fossilized footprints of animals and humans.

6. True.

7. No. Current observed conditions do not support slow fossilization over thousands, or millions, of years. Fossils cannot form slowly because organisms are rapidly scavenged and decay, which leaves no organism to be slowly covered by sediments. There is no organism left to slowly fossilize!

8. Strata are the layers that sediments form when they fall to the bottom of bodies of water. After the sediments harden and mineralize, strata are the layers that remain.

9. It is important to evolutionists because that helps to fit in with the theory that the earth is very old. If the strata formed quickly, then the earth could not be as old as evolutionists believe it is. If stratization was rapid, then there would be more strata than are currently seen to exist on earth. In addition, since the process of evolution is slow, the other associated processes must also be slow. It is important to creationists that strata can form rapidly because that helps them to explain their concept that the earth is young as well as the rapid stratization they believe occurred during the flood of Noah. Further, observed events concerning fossilization favor rapid, rather than slow, fossil (and strata) formation.

10. The basic concept is to measure the amount of a given radioactive isotope in a fossil or rock sample. Then, the starting amount of the isotope in the sample is estimated. The half life is finally determined by comparing the starting and remaining amounts of isotope. This allows one to calculate the number of half lives of the isotope that have passed. The number of half lives is then converted into an age for the sample.

11. The pitfalls include: over (or under)-estimation of the starting amount of a radioactive isotope in the sample; and, to assume that the only way that the amount of a radioactive isotope in a sample can decrease is by radioactive decay.

CHAPTER 18

1. A naturalistic philosophy means that nature, and nature alone, has determined what life forms currently are living, and have lived in the past. Another word for naturalistic philosphy is evolution.

2. It is of critical importance that the majority of evolutionary thought leaders are, and were, either atheist or agnostic. If a person does not believe in God, how could the universe and everything in it possibly have been created by God? Therefore, some other explantion must exist for the presence and diversity of life on earth. Evolution has fulfilled that explanatory need for the non-believer.

3. Charles Darwin was an English scientist in the 1800's. He is considered the father of evolution. His book, The Origin of Species, was the first cohesive thesis on the process of evolution.

4. Descent with modification is one of the fundamental principles of evolution. It describes the process in which less complex organisms are able to evolve into more complex ones over time (or the process in which more complex organisms evolve from less complex ones). Specifically, it refers to the traits of less complex organisms changing over time and being passed from generation to generation until a new, more complex species has been developed as a result of the passing of the new traits.

5. The most basic concept of natural selection is survival of the fittest. The organisms that are best adapted to their environment (or are more fit) are able to live and reproduce to pass on their traits for "survival fitness" to their offspring. Conversely, the organisms that have traits that make them less fit to survive in their environment will not live. From an evolutionary standpoint, natural selection is the process that drives descent with modification. From a creation standpoint, natural selection does occur, but does not lead to evolution. It only causes variation in traits within members of the same population of species.

6. Creationists definitely agree that natural selection occurs. It is readily observed to occur in nature studies. Creationists disagree with the evolutionary belief that natural selection leads to the formation of new species.

7. An adaptation is a specialized trait that allows an organism to survive better in its given environment.

8. A species is a specific type of organism. Or, a species is a group of organisms that can and do breed with one another in their natural habitat.

9. False. This is the description of convergent evolution. Coevolution is the evolution of traits of two unrelated species in close association with one another so that the adaptations of each (or the evolution of each) benefit the other.

10. The evolutionist views the adaptive radiation of the ancestor finches as evolution in action. As the ancestor finches reproduced and multiplied, they naturally inhabited different areas of the islands seeking food and nesting areas. These areas had unique food sources and habitat variations. As a result, the different populations would have developed different adaptations that allowed them to survive better in their given environment. Critical to this discussion is that the new adaptations (traits) occurred as a result of the ancestor finches gaining new genetic information coding for the new traits, passed from one generation to the next. Eventually, the populations gained enough genetic material, coding for new traits, that they became distinct and separate species from one another.

    The creationist views the adaptive radiation as natural selection in action. They believe the original ancestor finch population was created by God with great genetic diversity, which would give the potential for great trait variation depending upon the pressures created by natural selection. The traits of the finches would change slightly over time depending upon where they lived on the island and what their diet was. Over time, as a result of natural selection acting through their built-in genetic diversity, the 13 different populations developed. Critical to this view is that the genetic diversity was encoded into the finches (and into all species) by God when He created them. No finches gained genetic material encoding new traits. God created the process of natural selection and so He naturally created all organisms with the needed genetic diversity through which it could act.

11. According to evolutionary theory, there should be transition fossil forms; there are none. Lack of agreed-upon transitional forms leads one to believe that evolution does not occur and that, therefore, evolution cannot explain the origin of species. However, if a transitional form was to be found, then that would give credence to evolution being true.

12. Of course it is important that some evolutionary thought leaders do not believe there are transitional forms in the fossil record. First off, it is important that they, as scientists, show they are able to objectively assess the data and come to conclusions that do not necessarily support what they believe, but conclusions that are supported by fact. It also lends support to the creationist idea that evolution does not explain the presence of life on earth.

13. Creationists do not believe that vestigial structures represent evidence that evolution is true mainly because almost all of the structures that evolutionists once considered vestigial in humans have been found to have vital functions. Further, creationists believe that the few structures that have not been found to have an apparent function will some day be found to have a critical function as we study the human body more completely. Homology is considered invalid by creationists because they believe that God knew the best way to design all structures. He stuck with the best way to design them, which is why many structures that have the same function look the same. It is also why similar biochemical reactions have similar enzymes and proteins.

CHAPTER 19

1. Neo-Darwinsim, also called the modern evolutionary synthesis, is the attempt to explain evolution on a genetic level. The problem with neo-Darwinism is that the process upon which it rests - information-adding (gene-adding) mutations - has never been proven to occur. All mutations have been proven to remove information (by damaging the genes in which they occur), not add it.

2. Adaptations are traits that allow organisms to survive better in their environment. There are three general types of adaptations: behavioral, structural, and physiological.

3. False. The bright coloration is an example of a structural adaptation and the poison it produces is an example of a physiological adaptation.

4. True.

5. The changes in the finches observed by Dr. Grant was absolutely not evolution in action. By definition, evolution is the development of a new, more complex species with new traits from an existing, less complex species with fewer traits. In order for evolution to occur, new genes need to be added to the ancestor species that cause it to develop new traits and evolve into the more complex species.

   In the Galapagos finch population, there was neither the addition of genetic material to the finches, nor was there the formation of new species. Since the criteria for evolution were not met, the changes observed in the finch population was natural selection in action, not evolution.

6. According to evolutionary doctrine, speciation occurs as the result of natural selection, leading to descent with modification. Critical to this process is the addition of genes, coding for new traits, through a genetic process that has never been proven to exist. According to creation doctrine, every species was created by God with a large amount of genetic variability. That genetic variability allowed for normal trait variation, through which natural selection could act. This process led to the currently present genetic diversity.

7. Paleoanthropologists are scientists who study human origins and evolution.

8. According to evolutionists, the first hominid was one of the australopithecine variants called A. ramidus, arising 4.5 million years ago.

9. Please review the answer to this question with your parents.

CHAPTER 20

1. First, it provides a very well-described system to classify all organisms. Second, it allows all organisms to be classified in such a way that there is no confusion regarding organisms being studied or discussed.

2. There were several. There was not the ability to properly accomodate the huge number of organisms that have been, and are yet to be, discovered and classified. The older system was not specific enough, which often led to confusion. In addition, the older systems relied mainly upon the use of the common local names for organisms, further adding to the confusion, as many organisms have more than one common name.

3. Seven.

4. False. Referring to an organism using the binomial name completely elimintaes any confusion regarding exactly which organism is being referred to since there is only one organism with a given binomial name.

5. The two main criteria are structural and biochemical (or physiological) similarities and differences.

6. Escherichia coli, *Escherichia coli*, E. coli, and *E. coli* are all correct.

7. False. The Kingdom level is the least specific level of classification. The species level is the most specific.

8. All prokaryotes are classified into one of two Kingdoms - Archaebacteria and Eubacteria.

9. All eukaryotes are classified into one of four Kingdoms - Protista, Fungi, Plantae, and Animalia.

10. There are several reasons viruses are not considered living organisms, and include the facts that they: do not reproduce on their own, cannot maintain homeostasis, do not have a metabolism, and they do not contain both DNA and RNA

11. See Figure 20.11.3.

12. A DNA virus is a virus that contains DNA as its nucleic acid.

13. In the lytic cycle for a DNA virus, the DNA is injected into the host's cytoplasm. It then immediately commandeers the host's internal machinery to begin making more viral components. Viral DNA is transcribed into viral RNA, and viral RNA is translated into viral proteins. The host cell makes virus particles in this way until it literally explodes with new viruses.

    In the lysogenic cycle, the viral DNA integrates (inserts) into the host DNA, rather than commandeer the host's machinery. It then remains in the host DNA for a time, replicating every time the host cell replicates. Then, when it is stimulated, the virus enters the lytic cycle.

14. Retroviruses were key to early DNA technology because they contain the enzyme reverse transcriptase. This is the enzyme that allows geneticists to work backwards in DNA technology.

15. False. Like viruses, viroids and prions are not considered to be living organisms.

CHAPTER 21

1. Monera is the kingdom into which all prokaryotic species were classified. Now these organisms are classified into Archaebacteria or Eubacteria.

2. The three most common bacterial shapes are rod-shaped (bacillus), spherical-shaped (coccus), and spiral-shaped.

3. See Figure 21.4.1.

4. False. All bacteria have cell walls.

5. False. Gram-staining is a procedure used to identify Eubacteria. Archaebacteria are usually not identified based upon the Gram-stain.

6. Gram-positive bacteria stain "positive" because they have a very thick peptidoglycan layer in their cell wall. This thick layer traps the dark blue/purple colored crystal violet stain that is applied during the first step of the Gram-stain. As such, it is not washed out by the alcohol. After the Gram-stain procedure is complete, Gram-positive bacteria appear dark blue or purple under the microscope.

   Gram-negative bacteria stain "negative" because they have a thin peptidoglycan layer in the cell wall. This thin layer does not trap the crystal violet stain and so it is washed out by the alcohol during the alcohol wash step of the Gram-stain procedure. This causes the Gram-negative bacteria to lose the purple color. In the last step of the Gram-stain procedure, the red/pink colored dye safranin is added, which stains the Gram-negative bacteria pink/red, and so they appear pink or red under the microscope.

7. All antibiotics take advantage of the fact there are different metabolic processes and different enzymes in prokaryotic bacteria as compared to their eukaryotic hosts. Antibiotics target and inhibit the unique processes/enzymes of bacteria, while not harming the eukaryotic host. This causes antibiotics to kill bacteria but leave the host unharmed.

8. True.

9. Archaebacteria are known for their ability to live and thrive in extreme environments.

10. When the term "anaerobic" is applied to bacteria, it means that a certain bacterial species has the ability to grow/survive without oxygen. Obligate anaerobes need to grow in the absence of oxygen; that is, oxygen is toxic to (or kills) obligate anaerobic bacteria. Facultative anaerobes are able to grow in the presence or absence of oxygen. But, facultative anaerobes do not need oxygen to live.

11. False. Photosynthetic bacteria contain chlorophyll but not chloroplasts.

12. An endospore is a hardy bacterial unit that forms during times of environmental stress. Endospores contain bacterial DNA and have the ability to grow (or develop) into a new bacterium when conditions are favorable for growth.

13. Endotoxins are part of the outer membrane of Gram-negative bacterial cell walls. Exotoxins are proteins secreted by Gram-positive bacteria. Both are harmful when they are released into the host.

14. The cell wall structures of Archaebacteria and Eubacteria are very different. Also, the amino acid sequences of tRNA and rRNA are very different between the members of the two kingdoms.

CHAPTER 22

1. The three classes are protozoa (also called the animal-like protists or heterotrophic protists), the algae (also called the plant-like protists or autotrophic protists), and the slime molds/water molds (also called the fungus-like protists or absorptive protists).

2. Plankton forms the bottom layer of the food chain in bodies of water from which all other organisms ultimately derive their energy and carbon sources. Also, the photosynthetic plankton (phytoplankton) produce the majority of the atmospheric oxygen. Protists are important in relation to plankton because they make up a large percentage of the zooplankton and phytoplankton layers.

3. True.

4. True.

5. See Figure 22.3.1.

6. See Figure 22.5.3 for the reproductive cycle. Alternation of generations means that an organism exists in the haploid gametophyte stage in one generation and then reproduces and grows into a diploid sporophyte the next generation, which then reproduces and grows into the haploid gametophyte generation, and so on.

7. False. Not all algae are unicellular organisms. In fact, it is some of the multicellular algal species that are the tallest organisms on earth.

8. True.

9. Harsh environmental conditions often trigger protists to enter into sexual reproduction.

10. Amoeboid movement occurs as a result of cytoplasmic streaming, forming pseudopods.

11. Conjugation (as discussed in this chapter) represents the exchange of genetic material between two paramecia.

12. Sporozoans are studied so extensively because they cause more human infections than any other type of organism on the planet.

13. They obtain their energy by breaking down dead and decaying matter. They are important decomposers.

CHAPTER 23

1. False. No fungal organisms are photosynthetic. They are all heterotrophs.

2. All Fungi have cell walls made of the polysaccharide polymer chitin.

3. All Fungi are heterotrophic. They obtain their energy through extracellular digestion.

4. A hyphus is the basic building block of a fungus.

5. Mycelia are formed by hyphae. A mycelium is the growing part of a fungus. It is the part of the fungus that performs extracellular digestion and then absorbs and transports the nutrients to the rest of the fungi (including the rest of the mycelium as well as the fruiting body).

6. False. Most Fungi exist in the haploid state during their life cycle.

7. False. In general, Fungi are able to reproduce sexually and asexually.

8. The three commonly accepted divisions (phyla) are Zygomycota, Basidiomycota and Ascomycota. However, in this course we used the four divisions of Ascomycota, Zygomycota, Basidiomycota, and Deuteromycota.

9. Fungi are broken into their divisions (phyla) based upon the ways in which they perform sexual reproduction and upon the sexual reproductive structures they form.

10. See Figure 23.6.1.

11. This description fits the organisms of Zygomycota.

12. True.

13. Antheridia are features of organisms classified into Acsomycota.

14. A lichen is a symbiotic relationship between a fungus and a photosynthetic bacterial species or an algal species.

15. Lichens are classified into the proper seven levels based upon the fungal component of the lichen (i.e. they are named and classified as the name of the fungal species).

16. Mycorrhizae is the term used to describe the relationship of a fungus growing in and around plant roots. This generally denotes a symbiotic relationship between the two members of the mycorrhizae.

17. False. Many fungal species are harmful to humans, plants, and animals, some are even fatal.

CHAPTER 24

1. A scientist who studies plants is called a botanist.

2. True.

3. False. All plant cell walls are made from cellulose.

4. True.

5. As you walk from the inside of the nucleus to the outside of the cell, you would pass through the following structures in this order: nucleoplasm, nuclear membrane, cytoplasm, cell membrane, secondary cell wall, and finally the primary cell wall.

6. Lignin is a chemical compound that is deposited in the secondary cell walls of plants that have woody stems. It is beneficial because it gives the plant extra stability and rigidity.

7. The easiest way to move from one plant cell to another is to travel through the plasmodesmata.

8. Xylem and phloem are specialized plant tissues. They are found throughout the plant in the roots, stems, and leaves. They are very efficient tissues at conducting minerals, water, and food from one part of the plant to another. Plants that contain xylem and phloem are called vascular plants.

9. Seeds are specialized plant reproductive structures. They contain endosperm and the plant embryo and are covered by the seed coat. Not all plants form seeds. The two groups that do are called the gymnosperms and angiosperms.

10. There are several differences between monocots and dicots but the main difference is in the seed structure. Monocots form seeds with one cotyledon, and dicots form seeds that have two cotyledons. Together, the monocots and dicots form the larger group of plants called angiosperms.

11. Nonvascular plants are plants that do not contain xylem and phloem.

12. Rhizoids are not true roots because they do not contain xylem and phloem. Botanists have defined "roots" as plant structures that contain xylem and phloem and absorb water and nutrients.

13. A taproot system.

14. All plant growth occurs at the meristem tissue.

15. False. Plants demonstrate high degrees of organization in their root, stem, and leaf structures.

16. Tree (or growth) rings form as a result in differences in growth rates during the spring and early summer as compared to the late summer and fall. In the spring and early summer, water and nutrients are plentiful which allows the tree to grow faster. As a result, the xylem cells are very large in order to accommodate the transport of large quantities of water and nutrients. This results in a wide layer (ring) of light-colored xylem forming in the beginning of the growth season. Later in the year, the growth slows and the xylem cells are smaller, causing the formation of a dark-colored thin ring of xylem. During periods of wetness with good nutrition, growth rings are thicker (wider) than they are during dry periods.

17. Guard cells open and close leaf stoma. They are found in the lower epidermis, on the underside of the leaf, surrounding every stoma. They open and close the stoma depending upon the presence or absence of water. When water is plentiful the stoma are open so that photosynthesis can occur. When water is scarce, the guard cells close the stoma, causing photosynthesis to stop, which helps the plant conserve water.

18. Photosynthesis occurs in the palisade mesophyll.

19. Turgor pressure is pressure that exists due to water. Turgor pressure normally exists inside every plant cell, and every plant. When water is plentiful, all of a plant's cells are filled with water, which results in the formation of an outward pressure of one cell to another so that all of the cells push against one another. This is turgor pressure and causes plants to be able to maintain the proper "posture" (i.e. not wilt) so they can perform photosynthesis and grow properly.

20. True.

21. The cohesion-tension theory describes the process of moving materials through xylem. This theory is akin to sucking liquid out of a cup using a straw. Cohesion describes the property of water molecules wanting to always stick together as the result of hydrogen bonding. The process begins with transpiration. As water escapes from the leaves, it results in the formation of a "pulling" force that serves to draw more water and materials up the plant through the xylem to replace the water lost from the leaves. This pulling force is caused due to the cohesion of water. This results in the constant movement of material from the roots, up the stem, and toward the leaves.

22. Transpiration is the loss of water through plant leaves. It is basically evaporation that happens to occur through (or out of) the leaves of plants. It is a critical process in that it is necessary for movement of materials through the xylem as we just saw in the description of the cohesion-tension theory.

CHAPTER 25

1. Hormones are chemical compounds that are produced in one tissue type and released into the organism to affect the function of another tissue. Plant hormones affect the physiology of the plant. The six hormones we discussed are: auxins, gibberellins, abscissic acid, ethylene, cytokinins, and florigen.

2. True.

3. There are two basic differences. The first relates to how the movements occur. Tropisms occur as the result of the effects of plant hormones. Nastic movements occur as a result of changes in turgor pressure. The second difference is that tropisms occur due to the direct response to a specific stimulus while nastic movements are the result of a non-specific stimulus.

4. Ethylene is a plant hormone that causes fruit to ripen. It is commonly used to ripen fruit after the fruit has been shipped in an unripened condition. Unripened fruit is hardier and travels better than ripened fruit.

5. Vegetative reproduction is a form of asexual plant reproduction. It is the formation (or growth) of a new plant from a part of a plant that is normally non-reproductive. Virtually every plant is able to reproduce through vegetative reproduction.

6. Alternation of generations is a reproductive cycle that all plants exhibit. It refers to the fact that plants exist in one phase of their life cycle as a diploid sporophyte that reproduces with spores. Then the spore grows into the next generation, which is a haploid gametophyte that reproduces with gametes. A male and female gamete then combine to form the next generation - the sporophyte.

7. True.

8. The gamteophyte phase is the dominant form (phase) of most nonvascular plants.

9. Chances are you would be looking at the gametophyte phase of a fern.

10. Sori are on the undersurface of fronds and contain spores.

11. The contents of a cone depend upon whether it is a male or female cone as well as where in the reproductive cycle the cone is. A female cone may contain female spores, the female gametophyte, or seeds before they are released. The male cone may contain the male spores or it may contain the male gametophyte (pollen) before it is released from the cone. Female cones tend to be on the higher branches, and they tend to remain on the branches longer as the male cones fall off the tree after they release pollen. Male cones usually form in bunches on the lower branches, near the trunk. Male cones are smaller than the female cones.

12. See Figure 25.9.2.

13. Nectar and bright coloration are flower traits that help to attract animals to pollinate the flowers.

14. The anther produces the male spore, called the microspore mother cell. This then develops into the male gametophyte (pollen).

15. Following pollination in flowering plants, a seed is formed, around which a fruit develops.

16. Germination is the process of a plant embryo breaking through the seed coat and beginning to grow into a new plant.

17. See Figures 25.9.3, 25.9.4, and 25.10.1.

18. The largest group of plants is the angiosperms.

19. A gametophyte is a haploid plant that reproduces with gametes. Gametes are also haploid.

20. A sporophyte is a diploid plant that reproduces using a spore. Spores are haploid.

CHAPTER 26

1. Animals are classified as vertebrate and invertebrate.

2. All animals share the following characteristics: eukaryotic cell structure, no cell walls, multicellularity, extracellular matrix, tissues, sexual reproduction, mobile, heterotrophic, energy storage in the form of glycogen, and aerobic metabolism.

3. Classification criteria include: body symmetry, embryonic tissue layers and development patterns, segmentation, the presence and type of body cavity, physical appearance, and biochemistry.

4. A jellyfish has radial symmetry, giraffes and snakes have bilateral symmetry (all vertebrates do). A starfish has a special type of radial symmetry called pentaradial symmetry.

5. You sit on the dorsal surface of a horse. A snake crawls on its ventral surface.

6. Cephalization is a term that describes the concentration of sensory tissue (brain, ears eyes, nose) in the anterior (head) of an animal.

7. True.

8. You would be looking at Cnidaria or Ctenophora.

9. Morphogenesis describes the process of embryonic development in animals.

10. The balstula forms before the gastrula during morphogenesis.

11. False. The skeleton and muscle are derived from mesoderm but the brain is derived from ectoderm.

12. False. The notochord develops into vertebrae. The neural tube develops into the brain and spinal cord.

13. Animals that have direct development have an immature stage that resembles a smaller version of the adult. As the animal grows, it gradually takes on the characteristics of the adult. Animals with indirect development have an immature (larval) stage in which the organism looks nothing like the adult form. The larva undergoes metamorphosis, during which time it takes on the characteristic appearance of the adult form.

14. True.

15. Invertebrates can have one of two support structures - the water or an exoskeleton. The support structure of a vertebrate is called an endoskeleton.

16. Invertebrate nervous systems are fairly simple. Vertebrate nervous systems are highly complex and regulated structures. Invertebrates can reproduce in a number of different ways. All vertebrates reproduce sexually, and have direct development (usually). Invertebrate digestive "systems" can include absorbing the food directly from their surroundings or through a gut. All vertebrate digestive sysytems are with a well-formed, multi-organ gut. The excretory systems of invertebrates can include ridding the wastes directly from the cells to the environment or through simple tissues. Vertebrate excretory systems are with multiple organs, including kidneys, that filter the blood and remove the wastes.

    Invertebrate support is through the water or an exoskeleton while vertebrate support is through an endoskeleton. Invertebrates can either have no circulatory system, an open system, or a closed system. All vertebrates have a multi-organ, closed circulatory system. Invertebrate gas exchange is done directly with the environment or with gills. Vertebrates either have gills or lungs.

17. True.

18. A gemmule is a small bud of sponge tissue that has the ability to grow into a new sponge organism. This is a form of asexual reproduction that sponges perform in harsh environmental conditions.

CHAPTER 27

1. Cnidarians demonstrate radial symmetry.

2. See Figure 27.2.1.

3. A cnidocyte is a specialized stinging cell that is only found in organisms classified in Cnidaria. Cnidocytes contain a special organelle called a nematocyst.

4. Platyhelminthes is the most "primitive" phylum that demonstrates cephalization.

5. Cnidaria and Platyhelminthes have two-way digestive tracts.

6. The organisms of Nemertea have a simple, but complete, one-way digestive tract.

7. Nemertea is the first phylum in which a completely formed vascular system is first seen.

8. Organisms of Platyhelminthes and Nematoda are common human parasites.

9. Ganglia are small collections of nerve cells. They are found in mollusks (as well as many other types of organisms). Ganglia are interconnected and work to coordinate an animal's reception and response to changes in the environment.

10. True.

11. Mollusks have an open circulatory system.

12. A cerebral ganglion refers to a collection of nerve tissue found in the anterior region (head), which functions like a brain. This is found in organisms of Annelida.

13. False. Annelids do not have lungs.

14. The scolex is the head of a tapeworm, found in organisms of Platyhelminthes. The scolex is designed to securely attach the tapeworm onto the inside surface of the host's intestines.

15. Nephridia are specialized excretory tissues found in annelids. They remove harmful wastes, such as ammonia, from the organism.

16. This is kind of a hard question to definitively answer, depending on what you consider to be an appendage. Acceptable answers include Cnidaria, Mollusca, or Arthropoda. One could make the argument that the tentacles of cnidarians or mollusks are properly referred to as appendages. However, Arthropoda is the first phylum in which jointed appendages are seen.

17. Arthropods have a rigid exoskeleton made of chitin. This causes problems in regards to growth, because a rigid outer skeleton does not "give" during growth. In order to overcome this situation, the exoskeleton is shed several times in the arthropod's life, a process called molting.

18. Organisms that undergo incomplete metamorphosis do not show as severe of a change in their body structure pre- and post-metamorphosis as compared to those organisms that undergo complete metamorphosis. In incomplete metamorphosis, the immature form of the organism is called a nymph, which looks like a smaller version of the adult, but it is lacking some of the adult structures and it is not sexually mature. A nymph molts several times, and after the last molting it looks like the mature adult. Organisms that undergo complete metamorphosis emerge from the embryonic stage in a very immature form, called the larval stage, that looks nothing like the adult form. The larva undergoes several moltings, and then enters the pupa stage. As a pupa, it completes the metamorphosis and emerges as the mature adult form.

19. Aquatic arthropods use gills for gas exchange. Terrestrial arthropods have a series of interconnected air tubes, called tracheae, that bring air into the animal. Then the cells individually exchange gas with the air in the tracheae.

20. Organisms of Chelicerata (spiders) have book lungs.

21. An endoskeleton is first seen in the organisms of Echinodermata.

22. False. They may move slowly, but sea stars and sand dollars are mobile.

23. Grouping the chordates into "vertebrate" and "invertebrate" is a convenient way to sub-divide this phylum.

24. All chordates have a notochord and a dorsal, hollow nerve tube at some point in their lives.

CHAPTER 28

1. The key feature that all organisms in Vertebrata share is a dorsal spinal cord surrounded by vertebrae (the spinal column).

2. First off, it is critically important to understand that neo-Darwinism has never actually been proven to occur. A second problem is that there are no fossil intermediates and all organisms appear in the fossil record as fully formed, looking then as they do now. In addition, there is no fossil evidence of changes in organisms that evolutionists believe to have been present on earth for several hundreds of millions of years. For example, when the fossils of sharks and crocodiles are examined, they demonstrate no change in their structure even though they are supposedly several hundred million years old. It also seems very improbable that birds (what evolutionists refer to as the dinosaur's nearest living relative) took more than 150 million years longer to evolve from reptiles, as compared to mammals.

3. An organ is a group of tissues that perform a common function.

4. A lateral line is found in organisms of Chondrichthyes and Osteichthyes. It runs up and down the sides of fish and brings information regarding the animal's environment to the brain. It is a highly sensitive sensory structure.

5. Organisms of Chondrichthyes have an endoskeleton made of cartilage, while those of Osteichthyes have an endoskeleton made of bone.

6. Aquatic, non-mammalian vertebrates exchange gasses using gills.

7. The atrium is the chamber that receives blood returning to the heart. The ventricle is the chamber that pumps blood out of the heart.

8. See Figure 28.6.4. This is a one-loop circulatory system of the fish.

9. Capillaries are specialized blood vessels that perform gas, nutrient, and waste exchange.

10. In order to answer this, think of what amphibians have that their supposed ancestors, the fish, don't have, and also think of what amphibians don't have that fish do. For example, amphibians have lungs for gas exchange, fish have gills. Also, amphibians have a two-loop curculatory system, fish have a one-loop system. It is easy to explain the loss of a structure on a genetic basis because of what mutation is known to do. Mutations cause a loss of information within a gene, resulting in a loss of traits. Therefore, it is not hard to explain how a fish ancestor would have lost its lungs during the evolution to an amphibian - it would have had a mutation in the genes coding for the production of gills. However, this is only part of the story.

    At the same time the amphibian ancestor fish was losing its gills, it would have also had to be gaining meaningful genetic information through mutation (again, a process that is known not to occur) that coded for the production of lungs. In addition, at exactly the same time, the same ancestor fish species would have had to be gaining genes that coded for a change in the circulatory system from a one-loop to a two-loop system AND it would have had to be gaining information that caused it to grow legs (lungs are not much good if you are not able to breathe air).

11. A tadpole is the larval stage of an amphibian.

12. See Figure 28.7.4. This is a two-loop circulatory system.

13. False. All land vertebrates have a two-loop circulatory system as adults.

14. Birds have more heart chambers than amphibians. Birds have a four-chambered heart, while amphibians have a three-chambered heart.

15. The blood does not mix in amphibians because of the way the atria contract and the way the internal anatomy of the ventricle is designed. These two factors direct the blood in a fashion that does not allow the oxygenated and deoxygenated blood to mix together even though they are present at the same time in the same single ventricle.

16. A nictitating membrane covers the eyes in some aquatic vertebrate species. It is a clear membrane, and so when the organism is under water, it closes to protect the eye, but still allows the organism to see since it is clear.

17. Amphibians are dependent upon water because their eggs are soft and do not have a protective covering. They must be laid in water to avoid drying out (which kills the embryo). Also, amphibian skin must be kept moist so it does not dry out and so proper gas exchange can occur across the skin. Reptiles have amniotic eggs, which provide a moist environment for the embryo while protecting it with a hard shell. Reptile skin is very tough and water-tight and does not need to be kept moist with water.

18. See Figure 28.8.3. This is an important structure to land based animals because it allows them to be independent of water for the survival of the embryos.

19. Jacobsen's organ is specialized reptile tissue found in the mouths of snakes, lizards, and turtles.

20. An organism that is oviparous is fertilized internally, and then a shell forms around the embryo. The female then lays the egg and the embryo completes its development and then hatches outside of the female. An ovoviviparous organism is internally fertilized, and then a shell forms around the embryo. However, the egg remains in the female (meaning it is not laid) until it hatches, or until just before it hatches. An organism that is viviparous is internally fertilized and forms an embryo that does not have a shell around it. The embryo is retained in the mother until it is developed and live birth is given.

21. An endotherm is able to produce its own body heat, a condition called "warm-blooded." Endotherms perform many more chemical reactions than ectotherms and are able to maintain temperature homeostasis independent of the environmental temperature. Ectotherms cannnot maintain temperature homeostasis independent of the environment, a condition referred to as "cold-blooded."

CHAPTER 29

1. An ornithologist is a scientist who studies birds.

2. Birds have beaks, feathers, and air sacs.

3. The idea that birds evolved from dinosaurs is a long-standing evolutionary principle. Virtually every evolutionary book and all media sources promote this as "fact." Therefore, changing that thought would require a major paradigm shift on the evolutionist's part. This would result in great confusion since the evolution of reptiles to birds is largely promoted as the most definite evidence that evolution happens. It is much easier and consistent to promote the idea that birds evolved from dinosaurs even though many prominent evolutionists now do not believe that to be true.

4. Air sacs are air-filled spaces found in the chest and abdomen of a bird. They are continuous with the respiratory system. Airs sacs help to make the bird's body lighter than it would be if the air sacs were not present. This facilitates flight. Air sacs also provide a continuous flow of oxygen to the lungs so gas exchange can occur during both inhalation and exhalation. When the bird inhales, fresh air fills the lungs and the air sacs. The lungs perform gas exchange and the air sacs store the fresh air. Then the bird exhales, releasing the air in the lungs (that is high in carbon dioxide and low in oxygen) into the air. Fresh air from the air sacs flows into the lungs and gas exchange continues during exhalation.

5. True.

6. Some birds navigate using the sun, others use the stars, and still others use the Earth's magnetic field.

7. True.

8. The keel is the enlarged breastbone of a bird that serves as the attachment site for the powerful flight muscles.

9. Birds have specialized, light-weight hollow bones and air sacs, both of which make their bodies light.

10. The main difference is in the way they reproduce/bear young. Monotremes reproduce with eggs. Marsupials give birth to very immature young that need to complete development in the pouch. Placental mammals develop a placenta that allows the mother to carry the baby until it is completely developed.

11. True.

12. False. All mammals have a vertebral column.

13. True.

14. False. All mammals have a single lower jaw bone called a mandible.

15. False. All mammals perform gas exchange with lungs.

16. False. Mammal tails, properly called "flukes," contain no bones. They contain muscle and fibrous tissue.

17. The underlying concept behind all evolutionary principle is that life evolves in a constant move "forward" and should not move "backward." The movement of aquatic animals to land animals is considered a major evolutionary move forward. This means there is a move forward with aquatic animals developing more complex traits as a result of natural selections and it is these complex traits that caused aquatic animals to develop the more complex traits they needed to move out of the water. One of the major evolutionary steps was the movement of animals to land. Evolutionists believe that aquatic mammals evolved from terrestrial mammals, which is, evolutionarily speaking, a move backwards (which is something that should not happen as a result of natural selection).

18. False. The legs of aquatic mammals are small, but they have function and therefore are not vestigial. They support the pelvic organs.

19. Echolocation is the use of soundwaves to locate objects in the animal's environment. The mammalian orders of Cetacea and Chiroptera are capable of echolocation.

20. The area of the brain in question is the cerebrum. It is responsible for thinking.

21. An ungulate is a hoofed mammal.

22. False. Not all mammals are viviparous. Monotremes are ovoviparous.

23. False. Human and mammal organ systems function in much the same way.

CHAPTER 30

1. The nervous system is a specialized collection of tissues and organs that senses changes in the environment and transmits the sensory information to the super-controller of the nervous system - the brain. Then the brain makes a decision about what to do as a result of the sensory information. The decision results in a change in the organism's behavior, and the information to change to the behavior is transmitted from the brain through other structures in the nervous system to enact the change.

2. Answers will vary depending upon how you interpreted the question. One way to answer this is the nervous system is composed of the central nervous system, the peripheral nervous system, and the autonomic nervous system. The other acceptable answer is the nervous system is composed of the brain, spinal cord, spinal nerves, and peripheral nerves.

3. The two types of nervous system cells are neurons (nerve cells) and glial cells (support cells or astrocytes). Neurons conduct electrical impulses and glial cells provide the structure and support for the neurons to associate with one another and form the nervous system.

4. See Figure 30.3.2.

5. Sensory nerves bring information into the brain from the environment and body. Motor nerves carry information from the brain to the muscles to make them move.

6. skin ⟶ sensory neuron ⟶ peripheral nerve ⟶ spinal nerve ⟶ spinal cord ⟶ brain

   spinal cord ⟶ spinal nerve ⟶ peripheral nerve ⟶ motor nerve ⟶ motor neuron ⟶ muscle

7. The event that causes a sensory nerve to fire (or generate an impulse) is called a stimulus.

8. False. At rest, a neuron is polarized. When a neuron depolarizes, it conducts an electrical impulse, which means the nerve is no longer at rest, but is active.

9. See Figure 30.5.1. This depolarization event is called an action potential.

10. The nerve repolarizes when the Na-K-ATPase pumps turn back on, pumping sodium out of the cell and potassium into the cell.

11. False. Nerves do show an all or none response. Anytime the stimulus threshold is reached, the nerve will fire an impulse.

12. True.

13. The Na-K-ATPase pump is the protein that pumps sodium out of and potassium into the neuron. This is an active process that requires energy, which is supplied by ATP. This transmembrane protein is responsible for maintaining the resting potential of the neuron, as well as for reestablishing the resting potential following depolarization.

14. The pia mater, arachnoid, and dura mater are protective tissue coverings of the brain and spinal cord.

15. The CNS is composed of the brain and spinal cord. The PNS is composed of spinal nerves and peripheral nerves.

16. When an area of the brain in injured, the function of that area is lost or altered. The frontal lobes are concerned with thinking, initiating speech generation, and movement control. Therefore, if the frontal lobes are injured, a person would be expected to have problems thinking, moving, and speaking. Practically, a person may have trouble thinking into the future, balancing a checkbook, or speaking clearly. They may also have weakness on one side of the body such that they cannot walk, bathe, or dress normally.

17. The cerebellum controls our coordination and fine-tunes our muscle contractions. Damage to the cerebellum causes uncoordinated muscle movements. This causes problems walking, buttoning clothes, writing, and brushing teeth.

18. True.

19. True.

20. The white matter is composed of myelinated axons, and myelin is a white-colored substance.

21. Nerve cell bodies (neuron bodies) are contained in the grey matter.

22. The autonomic nervous system controls all of our automatic functions. It has two components - the sympathetic and the parasympathetic nervous systems.

23. The order that you would pass through those structures is: cornea, pupil, lens, vitreous, and retina.

24. False. Cones process color and rods sense light and dark (sometimes it is said that rods allow us to see black and white).

25. Sound waves first pass through the external ear, through the auditory canal, and then strike the tympanic membrane. The tympanic membrane vibrates when the sound hits it, which then vibrates the middle ear ossicles. The ossicles are connected to the oval window, a part of the inner ear/cochlea. As the oval window vibrates, the endolymph in the cochlea moves back and forth, which moves the cilia that are attached to the nerves of the auditory nerve. The movement of the cilia stimulates the auditory nerve cells to generate an impulse that is then conducted to the brain via the auditory nerve. In the brain, the impulses are sensed and processed to determine what the nature of the sound is.

26. See Figure 30.17.1.

27. Smell and taste receptors are stimulated by chemicals of different shapes or electrical charges. When a particular receptor on a particular cell is stimulated, it causes that cell to generate and send an electrical impulse to the brain for processing.

CHAPTER 31

1. Infectious organisms are referred to as pathogens.

2. The body's defense mechanism against pathogens is called the immune system.

3. The components of the immune system include physical barriers, chemical barriers, the nonspecific response, and the specific response.

4. Fevers are an elevation of body temperature in response to an infection. Fevers help the immune system fight off the pathogen more effectively.

5. The non-specific immune response includes histamine release, engulfing cells called macrophages and neutrophils, and fever.

6. The two branches of the specific immune response are cell mediated immunity and humoral immunity. Cell mediated immunity is a way to kill infected body cells using killer T-cells. Humoral immunity is a way to target and/or inactivate pathogens using antibodies.

7. Antigens are foreign proteins that stimulate an immune response.

8. The T helper cell is a type of T lymphocyte that presents antigen to the killer T cell and B cell so that the killer T cell can make receptors to the antigen and the B cell can make antibodies against the antigen. The killer T cell is a specialized T lymphocyte that can recognize and kill body cells that are infected with pathogens. Suppressor T cells are T lymphocytes that turn off the immune response once the pathogen is eliminated.

9. Immune system memory refers to the property of the immune system maintaining the long-term ability to recognize a previous pathogen. It remembers the invader and immediately removes it from the body.

10. An autoimmune disease is caused when a person's immune system loses the ability to recognize the person's body cells as "self." It is a disease that results when a person's immune system attacks their own tissues.

11. The organs and glands of the endocrine system communictae with one another using hormones.

12. The difference between these two types of hormones is how they are built. Amino acid derived hormones are built from amino acids. Steroid hormones are built upon a backbone of cholesterol. In addition, amino acid hormones are not fat soluble and therefore cannot pass through the cell membrane. Their receptors are located on the cell surface. Steroid hormones are able to freely pass through the cell membrane and their receptors are in the cytoplasm.

13. The hypothalamus controls the endocrine system.

14. The heart of the endocrine feedback loop system is the hypothalamus. The hypothalamus can sense levels of hormones made by all endocrine glands in the body by sampling blood. If the hypothalamus senses that an organ is not making enough of its hormone, the hypothalamus releases a hormone (or causes the pituitary to release a hormone) to cause that organ to increase its production of its hormone. On the other hand, if an organ is making too much of its hormone, the hypothalamus senses that. Then, the hypothalamus releases a hormone (or causes the pituitary to release a hormone) to cause that organ to decrease its production of its hormone.

15. A hormone is a chemical messenger made in one tissue (organ) and released into the blood to affect the function of another tissue (organ).

16. In order to answer this question, you need to look at Figure 31.9.3 to see what the effect of insulin is. When insulin is released into the blood, it causes the cells of the body to absorb glucose into the cell from the blood. So if you are a molecule of glucose in someone's blood and insulin is released, you will be taken up into one of the body cells.

17. When the hypothalamus releases TRH, it stimulates the pituitary to release TSH. TSH causes the thyroid to release thyroid hormone into the blood. So, the ultimate effect of TRH release from the hypothalamus is for the thyroid to release thyroid hormone.

18. The excretory system is composed of two kidneys, two ureters, one urinary bladder, and one urethra. Kidneys filter the waste from the blood while sparing water, ions, and nutrients. Ureters conduct the urine from the kidneys to the urinary bladder, which stores the urine. The urethra conducts the urine out of the bladder and the body.

19. See Figure 31.10.3b.

20. This description fits an unhealthy kidney.

21. Dialysis is an artifical way to filter the blood and remove waste products from a person who has kidney failure.

CHAPTER 32

1. True.

2. The heart pumps blood through the circulatory system. Veins carry blood to the heart and arteries carry blood away from the heart. Capillaries are the vessels where gas, nutrient, and waste exchange occurs.

3. The human circulatory system is a two-loop system.

4. See Figure 32.2.1.

5. You would pass through or by those structures in the following order: left atrium, mitral valve, left ventricle, aortic valve, aorta, iliac artery, femoral artery, systemic capillaries, femoral vein, iliac vein, inferior vena cava, right atrium, tricuspid valve, right ventricle, pulmonic valve, pulmonary artery, lung, lung capillaries, pulmonary veins, and then back to the left atrium.

6. There are four heart valves, and they prevent blood from flowing the wrong way (i.e. prevent the backflow of blood).

7. The electrical impulse for the heartbeat originates in the sino-atrial (SA) node.

8. The AV node collects the electrical impulse for the heartbeat from the atria and passes it to the ventricles. The AV node delays the passage of the electrical impulse slightly, which allows the ventricles to completely fill before they contract.

9. Systole and diastole are the two phases of the heartbeat.

10. Blood pressure is the pressure that forms in the arteries as a result of blood being pumped into them by the left ventricle.

11. Hypertension is another word for high blood pressure. If a person's systolic blood pressure is over 140 or the diastolic blood pressure is over 90, they are said to have hypertension (or to be hypertensive).

12. Blood is composed of plasma, white blood cells, red blood cells, and platelets.

13. Red blood cells contain a special iron-containing molecule called hemoglobin, which is capable of binding to oxygen. Hemoglobin binds oxygen in the capillaries of the lungs and carries the oxygen to the tissues, in the form of oxyhemoglobin. In the systemic capillaries, oxygen is released from the red blood cell to feed the tissues with needed oxygen.

14. The upper respiratory tract is composed of the: nose, mouth, nasal sinuses, pharynx, larynx, and trachea. The lower respiratory tract is made up of the bronchi, bronchioles, and alveoli.

15. A bronchiole is smaller in diameter than a bronchus.

16. During a swallow, muscles pull the throat and esophagus upward, closing the opening of the airway against the epiglottis.

17. During inhalation, the diaphragm receives the signal to contract from the phrenic nerve. The diaphragm contracts and moves downward, which enlarges the thoracic cavity and creates negative pressure. The lungs expand to fill the space in the chest cavity, and as the lungs expand, negative pressure is created inside the lungs. This functions like a vacuum and draws air into the lungs through the upper airways. Following inhalation, the diaphragm relaxes, allowing the normal elastic recoil of the lungs to pull the lungs into a deflated position, forcing the air out of the lungs.

CHAPTER 33

1. The digestive system breaks down and absorbs nutrients, as well as water, and then eliminates the wastes that are left over.

2. You will pass through those structures in the following order: mouth, pharynx, esophagus, stomach, duodenum, jejunum, ileum, colon, and rectum.

3. Peristalsis is the wave of muscular contractions that propels food through the GI tract.

4. Sphincters are muscular bands that prevent food from flowing the wrong way in the GI tract. Two sphincters we discussed are the esophageal sphincter and the pyloric sphincter.

5. Pepsinogen is the inactive form of the enzyme pepsin. Pepsin and hydrochloric acid are secreted into the stomach by the stomach cells when we eat. In the acidic environment of the stomach, pepsinogen is converted into pepsin, which can then begin to enzymatically digest protein.

6. Chyme.

7. The pancreas adds various enzymes, as well as sodium bicarbonate. The enzymes break down carbohydrates, proteins and fats, and the sodium bicarb neutralizes the acidic pH of the chyme. The gall bladder adds bile salts, which coats fats and makes them easier to absorb.

8. The jejunum and ileum absorb nutrients.

9. Villi and microvilli significantly increase the absorbtive surface area of the intestine.

10. Lacteals are small vessels inside the microvilli that absorb fats.

11. The colon absorbs water and packages the solid wastes for removal.

12. Ligaments hold bones together at the joints.

13. Tendons attach muscles to bones.

14. Osteoblasts are immature bone cells that secrete matrix, turning cartilage into bone.

15. Ossification is the process of a cartilage matrix being turned into a bone matrix (ossifying). It occurs when the osteoblasts secrete bone matrix into the cartilage, turning the cartilage into bone.

16. Growth takes place at the epiphiseal plates of bones.

17. A hinge joint allows movement in only one plane, but ball and socket joints allow movements in all planes.

18. The skull joints are fixed joints.

19. See Figure 33.9.2.

20. Whenever you bend a joint, you are flexing it.

21. The neuromuscular junction is the area where the motor axon ends and communicates with the muscle cell.

22. Acetyl choline is the neurotransmitter that is released from the nerve terminal to diffuse across the synapse and contact the muscle membrane. It is what tells the muscle to contract and is contained in vesicles in the axon brush. When the impulse reaches the axon terminus, the acetyl chlorine is released from the vesicles into the synapse.

23. True.

24. Follicles are found in the dermis.

CHAPTER 34

1. Ecology is the study of the relationship that living organisms have with one another and their environment.

2. The highest level of complexity is the community. The next highest level is the population, and the least complex unit is the individual.

3. The biotic mass is the trees, birds, insects, squirrel, fish, water lillies, snake and the frog. The abiotic mass is the rocks and mud, water, and the dead log. Microscopic organisms such as protists and bacteria are members of the biotic mass that are there but cannot be seen with the naked eye. Yes, this would be an ecosystem if you chose to define it as such.

4. True.

5. False. If the population density is increasing, there are more people living in a certain area than there were previously. This means the area is becoming more crowded.

6. The answers are emigration and higher death rates because both lead to a decrease in population.

7. The transfer of energy in an ecosystem is much more complex than the simple linear relationship imparted by a food chain. Food webs are much better at describing the intricate energy transfer relationships present in an ecosystem.

8. 90% of the available energy is lost when energy is transferred from a lower to the next higher trophic energy level.

9.

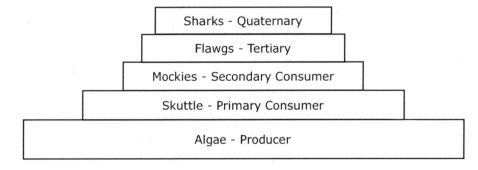

10. Camouflage specifically relates to the coloration and/or the shape of an organism that allows it to blend in with its environment. Mimicry specifically relates to the property of one animal looking like another such that a harmful organism physically looks like a non-harmful one, or such that a non-harmful organism resembles a harmful one.

11. Runoff is precipitation that has landed on the earth and runs along the earth's surface to be collected in a watershed. Percolation is precipitation that has struck the earth and been absorbed down into the earth to be stored in an aquifer (in the ground water). Both of these terms refer to components of the water cycle.

12. False. The greenhouse effect is not bad. It is a necessary phenomenon that causes the atmospheric air to be warmed enough so that earth can support life. Without the greenhouse effect, no life would exist on this planet.

13. True. IF global warming is true, then it will have disastrous consequences. The glaciers will all melt, releasing enough water to cover the entire surface of the earth several miles deep in water.

14. Fauna is the animal life of a biome and flora is the plant life of a biome.

15. Pioneer species are the first organisms that move into and grow in a new ecosystem.

16. Tundra contains mosses and lichens, as well as caribou, snowy owls, and artic foxes.Taiga contains evergreen trees with needles, moose, wolves, bears, and foxes. The temperate deciduous forest has broad-leaved deciduous trees/shrubs, white tailed deer and squirrels. The temperate grasslands contains long grasses and a few bushes, and animals such as prairie dogs. Desert contains succulents and animals such as snakes, lizards and small mammals. The savanna contains tall grasses with grazing animals and carnivores. Tropical rain forest has broad-leaved evergreen shrubs and trees that grow year round, and coloful birds and monkeys.

Biology Test Answer Key

Note to Parent: These tests are purposely designed to be in-depth knowledge assessment tools. As such, they are fairly long, so please do not be intimidated! Essay or short answer-type questions are the best way to identify whether or not your student completely understands the material. It is perfectly acceptable, if you choose, to have your student answer only the even questions, only the odd questions, or every third question as you see fit for your student's need. If they are able to correctly answer the majority of questions asked, it is likely they understand all of the material very well. If your student answers questions better orally, please ask them that way, rather than having them write it out. The method of test taking should not be a stumbling block in your student's performance.

BIOLOGY TEST #1 (Chapters 1-3)

1. Is the wood you burn in the fireplace alive? **No. Although it once was alive, as soon as it is cut down, the wood from the tree is dead.**

2. Why is asexual reproduction similar to making photocopies? **Because both asexual reproduction and making photocopies result in the production of identical objects. In the case of asexual reproduction, the offspring is identical to the parent. In the case of the photocopy, the copy is identical to the original.**

3. True or False? Although individual organisms are complex and organized, their interactions with one another and their surroundings are not. **False. Everything about organisms/ life is complex and organized.**

4. If an animal eats another animal to obtain its energy, what is that animal called? **A consumer, a carnivore or a predator are all okay answers.**

5. What is homeostasis? **The maintenance of a stable internal environment.**

6. What are the seven levels of classification? **Kingdom, Phylum (or Division for plants and fungi), Class, Order, Family, Genus, and Species.**

7. What is a binomial name? **It is the most specific level of classification for an organism. It is the Genus of the organism followed by the species. All binomial names are Latin.**

8. In an experiment, describe the difference between the experimental and the control groups. **The experimental group is exposed to the exact same conditions or factors as the control group except one. The factor that is different in the experimental group is called the variable. This type of experiment is designed to evaluate the effects of the variable on the organism.**

9. If new information is uncovered as a result of further experimentation that does not support a previous "theory," what must one do? **This situation occurs in science. Something that was previously thought to be "true," or was considered a theory, is proven not to be true. It is required that all scientists rethink the "theory" so that new information is taken into consideration. The theory must be retested and restated until all known information can be supported by the theory. If the theory cannot be supported by the new information, it must be thrown out.**

10. What is the difference between a multicellular and unicellular organism? **A multicellular organism is made of two or more cells and a unicellular organism is made of only one cell.**

11. Why are mass and matter not the same thing? **Matter is anything that takes up space. Mass is the amount of matter an object or area contains. Students often confuse the words "mass" and "weight" but technically, they do not mean the same thing. Weight is actually the amount of *force* exerted by an object. Weight is dependent on the force of gravity, but mass does not change. For example, if you**

took a bowling ball to the moon, it would weigh less on the moon than on earth. The mass of the bowling ball does not change. A bowling ball is a bowling ball. It contains the same amount of matter whether or not it is on the earth or the moon. But since there is more gravity on earth than on the moon, the bowling ball exerts more force (i.e. weighs more) on earth than on the moon.

12. What type of change in matter occurs when water changes into ice? **Physical.**

13. What type of matter change occurs when gasoline is burned in a car? **Chemical.**

14. What are the three subatomic particles that make up an atom? **Protons, neutrons, and electrons.** Which of these particles are in the nucleus? **Protons and neutrons.**

15. Why does an atom normally not have an electrical charge? **Because normally the number of protons equals the number of electrons in an atom, resulting in the atom having no overall electrical charge.**

16. What is the atomic mass of a hypothetical element with 27 electrons and 38 protons? **65. Add together 27 and 38.**

17. How many electrons can the first shell hold? **Two.** How many electrons can the third energy shell hold? **Eighteen.**

18. You have identified what may be two new elements. One element contains 212 protons, 212 electrons, and 97 neutrons. The other element contains 212 protons, 212 electrons, and 208 neutrons. Have you indeed found two new elements? **No. Since both forms of this element contain the same number of protons and electrons, they are the same element. However, you have discovered two new isotopes of the same element, since they contain different numbers of neutrons.**

19. What happens when a chemical bond is formed? **Two atoms share their electrons, which causes the two atoms to stick together.**

20. What are reactants and products? **Reactants are the atoms or molecules which "go into" (or participate in) a chemical reaction, and products are what "comes out of" (or results from) a chemical reaction.**

21. What makes one acid stronger than another? **The more hydrogen ions present in an acid solution, the stronger the acid will be.**

22. True or False? Organic molecules are synthesized by all living organisms. **True.**

23. What four types of molecules do organic chemists study? **Proteins, fats (lipids), carbohydrates, and nucleic acids.**

24. What is an isomer? **One or more molecules that share the same molecular formula but which have different structures.**

25. What is the difference between anabolic and catabolic reactions? **Anabolic reactions synthesize organic molecules; catabolic reactions break organic molecules down into smaller units.**

26. What is a monomer? **A monomer is the basic building block of all organic molecules.**

27. Why is a hydrolysis reaction so named? **Because water is used to break the bond between the monomer and the polymer. Since water is added to the bond, the name hydrolysis or hydration reaction is used.**

28. What is a glycosidic bond? **A glycosidic bond is a bond that links together two saccharides. It forms between a carbon of one saccharide, an oxygen molecule, and a carbon atom of the other saccharide.**

29. What is an ester bond? **A bond that holds a fatty acid to the glycerol backbone. Specifically, an ester bond is an oxygen atom bonded to a carbon on the glycerol and a carbon on the fatty acid, linking the two together.**

30. What is the monomeric unit of a protein? **An amino acid.**

31. What is an amino bond? **A bond holding two amino acids together.**

32. What are the three components of a nucleotide? **A central five-carbon sugar (called a pentose), a phosphate group, and a nitrogen-containing base (called a nitrogenous base).**

BIOLOGY TEST #2 (Chapters 4-5)

1. What is a group of cells called that all perform the same function? **A tissue.**

2. What is a group of tissues that all have the same function called? **An organ.**

3. What are the units inside cells that perform certain specific functions called? **Organelles.**

4. What features distinguish a eukaryotic cell? **A nucleus and membrane-bound organelles.**

5. If you were standing inside a cell and saw the DNA free-floating in the cytoplasm, in what type of cell would you be standing? **Prokaryotic. Since the DNA is free in the cytoplasm, you know you are not in a eukaryotic cell because the DNA of eukaryotic cells is contained in the nucleus.**

6. What is biosynthesis? **The process of making molecules out of absorbed molecules.**

7. True or False? Animal cells are protected from the environment by their cell membrane and cell wall. **False. Animal cells do not have cell walls.**

8. What is meant by the lipid bilayer? **The membrane of the cell forms into two layers of phospholipids because of the structure of the phospholipid molecule. The hydrophobic tails cluster together in the middle of the bilayer, and the hydrophilic heads protrude into the water matrix.**

9. What is meant by semi-permeable? **Semi-permeable refers to the fact that the cell membrane allows only certain substances to pass freely through. This effectively limits what goes into and out of the cell, protecting it.**

10. What is cellular transport? **The process of moving molecules/substances/"things" into or out of the cell across the cell membrane.**

11. What are two molecules that can pass through the cell membrane by diffusion? **Oxygen and carbon dioxide.**

12. What is the difference between active transport and passive transport? **Active transport requires the cell to use energy and passive transport does not.**

13. True or False? Gated channel transport, endocytosis, and exocytosis are all forms of active transport. **False. Gated channel transport is a form of passive transport.**

14. True or False? Diffusion occurs through the random movement of molecules down their concentration gradients. **True**

15. Describe how a protein pump works, and why it is considered active transport. **A protein pump pumps molecules up, or against their concentration gradients. This requires energy, so it is a form of active transport.**

16. Which cell type is structurally more complex: eukaryote, or prokaryote. Why? **Eukaryote. They have many more organelles than prokaryote cells.**

17. What is the cytosol, and what are its components? **It is the aqueous part of the cytoplasm. It is about 70% water and 30% ions and proteins.**

18. What functions does the cytoskeleton serve? **It maintains the shape (structure) of the cell. It also serves as an anchoring point for organelles and as the pathway for organelles and materials to be transported in the cell.**

19. How many lipid bilayers make up the nuclear membrane? **There are two lipid bilayers that make up the nuclear membrane. This means there are four individual lipid layers in the nuclear membrane.**

20. What is the structure and function of ER? **ER is a series of folded membranes that is connected directly to the nuclear membrane. ER functions to modify and transport proteins/molecules within the cell.**

21. What is the function of a vacuole? **Vacuoles function to store substances within the cell.**

22. Where is the middle lamella found and what is its function? **The middle lamella is found only in plants. It is a layer secreted outside the plant cell walls that holds plant cells together.**

23. If you were standing on the outside of a cell and found that the molecules around you were fibronectins, collagen, and proteoglycans, would you be standing in a plant or animal organism? **Animal. Plant cells are held together by sticky sugar molecules called polysaccharides.**

BIOLOGY TEST #3 (Chapters 6-8)

1. What is ATP? **The molecule almost all organisms use for energy is ATP; it provides the energy required to start endergonic reactions.**

2. What is metabolism? **The summation or totality of all chemical processes (or chemical reactions) occurring in an organism.**

3. What is the difference between chemical and mechanical energy? **Chemical potential energy is energy that is stored within chemical bonds of molecules. Chemical kinetic energy is work that is done as the result of the release of energy from a chemical reaction. Mechanical potential energy is energy stored in an object. Mechanical kinetic energy is energy that is transferred from one object to another.**

4. If energy is neither created nor destroyed in a system, where does the energy go when chemical-potential energy is converted into chemical-kinetic energy in an organism? **Most energy is lost as heat, the remainder of the energy provides the energy to drive the reaction.**

5. What is an enzyme? **There may be more than one answer for this depending on how far along the student is. Following this section, the answer is: an enzyme is a large protein molecule that regulates (controls) the chemical reactions occurring in a cell.**

6. Why does the cell use endergonic reactions? **Because they require an input of energy to start, which allows the cell to tightly control which reactions occur and when.**

7. What is the activation energy? **The energy required to start a chemical reaction.**

8. What is the induced-fit theory of enzyme-substrate binding? **The enzyme fits onto the substrate like a key fits into a lock. Once the enzyme binds the substrate at the active site, the enzyme then changes its shape slightly. The change in shape allows for the enzyme and the substrate to interact better during the chemical reaction.**

9. What is meant when enzymes are described as energy couplers? **Enzymes are able to add water to a phosphate bond of ATP. This releases a lot of energy. Enzymes are able to then take that energy and link it to the endergonic reaction, which the enzyme catalyzes. In doing so the enzyme couples the energy released from ATP to the reaction that the enzyme catalyzes so the reaction can proceed.**

10. How does feedback inhibition work? **When the level of a product of a reaction which an enzyme catalyzes gets too low in the cell, the enzyme is stimulated to "turn on." This results in the enzyme starting the reaction. The reaction then makes the product that is running low in the cell. As the reaction produces more and more of the product, the level of the product builds up inside the cell. Once the level of the product gets high enough, the level stimulates the enzyme to "turn off," and the reaction stops until the product level gets low again.**

11. What is carbon fixation and how does it relate to photosynthesis? **Carbon fixation is the process of incorporating carbon into organic molecules. Photosynthesis is the set of biochemical reactions that plants perform using the sun's energy to fixate carbon into organic molecules.**

12. How are heterotrophs and autotrophs related? **Autotrophs bring carbon into the organic molecules so both autotrophs and heterotrophs can use the organic molecules for energy. The organic molecules the autotrophs make are then used by heterotrophs to build their own organic molecules.**

13. How does radiant energy travel? **In discreet units called photons.**

14. True or False? The portion of the electromagnetic spectrum taken up by light represents the largest portion of the spectrum. **False. The portion of the spectrum taken up by light actually is the smallest portion of the electromagnetic spectrum.**

15. What portion of the electromagnetic spectrum provides the energy for photosynthesis to occur? **Photosynthesis is driven by the energy from the visible spectrum (or white light).**

16. When white light is passed through a prism, how does the light coming out of the prism differ from the light entering the prism and why? **A prism has the ability to refract light. Light entering a prism appears white. When it passes through a prism, the components of the white light are refracted (or bent) at different angles because they have different wavelengths. Light exiting the prism has been bent to different degrees, so all the components of the white light have been broken up. Light emerging from a prism looks like a rainbow, and has the main colors of red, orange, yellow, green, blue, indigo, and violet.**

17. What is the absorption spectrum of chlorophyll and why is it important to understand? **The absorption spectrum of chlorophyll includes all the wavelengths of light chlorophyll absorbs. The absorption spectrum of chlorophyll includes all colors except green and some yellow. (This means chlorophyll reflects green and some yellow wavelengths of light, which is why leaves look green). It is important to know this because the energy provided by the sun to fuel photosynthesis comes from the light chlorophyll absorbs.**

18. In which organelle does photosynthesis occur? **Chloroplasts.**

19. What is the overall purpose of the light reactions and where do they occur? **To release high-energy electrons from chlorophyll. This energy is then harnessed in the electron transport chain and used later in the Calvin cycle. They occur in the thylakoid membranes.**

20. Where are the enzymes of the light reactions contained? **The thylakoid membrane.**

21. Which photosystem is activated first? **Photosystem II.**

22. In the non-cyclic pathway, where are activated electrons obtained and how are they replaced? **They are obtained from chlorophyll. The electrons in chlorophyll are excited by the sun and leave the chlorophyll to enter the electron transport chain. The electrons are replaced from the splitting of the water molecule.**

23. Describe in detail the process of photosynthesis in the non-cyclic pathway. **In the thylakoid membrane, sunlight energy is absorbed by chlorophyll, generating activated electrons. The electrons leave chlorophyll and are replaced by electrons from water-splitting. Oxygen is released. The electrons are passed into the electron transport chain where their energy is captured to make ATP. The electrons are re-excited in PSI, then passed into another electron transport chain. This time, their energy is used to make NADPH. ATP and NADPH generated from the light reactions moves into the stroma, where the energy they contain is used in the Calvin cycle.**

24. How is ATP made in the electron transport chain? **As the electron is passed through the electron transport chain, it loses energy. At key points, the lost energy is harnessed by an enzyme in the electron transport chain and used to fuel the endergonic reaction, which makes ATP from ADP.**

25. Where is this ATP needed? **This ATP is then used for energy during the Calvin cycle.**

26. Where are the atoms for the carbohydrates made in the Calvin cycle obtained? **The carbon is obtained from carbon dioxide. Hydrogen and oxygen are obtained from water molecules.**

27. What are chlorophylls and carotenoids, and what do they do? **They are both photosynthetic pigments that absorb the energy contained in sunlight. In so doing, they start photosynthesis.**

28. Why is ATP made from glucose? **Glucose cannot be used directly by organisms for energy. Instead, the glucose is "burned," and ATP is made for the energy released from the bonds of the glucose.**

29. What is cellular respiration? **The process of breaking the bonds of glucose molecules and harnessing the energy released to make ATP.**

30. What is ATP used for? **It provides the direct energy needed for the cell to perform the endergonic reactions of life.**

31. What is the benefit of anaerobic respiration? **It generates ATP much more quickly than aerobic respiration.**

32. What is the benefit of aerobic respiration? **It generates much more ATP per molecule of glucose metabolized than anaerobic respiration does. This means it is more efficient.**

33. What are cristae? **Folds of the inner membrane.**

34. Where does the Krebs cycle occur? **The mitochondrial matrix.**

35. What happens during glycolysis and where does it occur? **Glucose is converted into two molecules of pyruvate in the cytoplasm.**

36. Describe the process of aerobic cellular respiration, starting with the transition reaction. Where does this occur? **Aerobic cellular respiration occurs inside mitochondria. During the transition reaction, pyruvate is attached to co-enzyme A (Co-A), forming acetyl Co-A. Acetyl Co-A then carries the acetyl group (two remaining carbon molecules from the pyruvate) through the Krebs cycle. During the Krebs cycle, ATP, NADH, and $FADH_2$ are produced. The NADH and $FADH_2$ then enter the electron transport chain. In the electron transport chain, energy carried in NADH and $FADH_2$ is converted into molecules of ATP.**

37. True or False? More ATP is generated in the electron transport chain than the Krebs cycle. **True.**

38. How many molecules of carbon dioxide are produced for every molecule of glucose that enters aerobic respiration? **Six. All six carbon molecules in glucose are metabolized to CO_2.**

39. Describe the proton motive force and chemiosmosis. **The proton motive force is the ability of hydrogen ions to perform work. It develops because proton pumps actively pump hydrogen ions across the cristae membrane. The hydrogen ions build up in high levels on one side of the cristae. As the protons (hydrogen ions) move back across the membrane as a result of gated diffusion, their movement across the membrane through the protein gate drives the enzyme ATP synthase. This makes ATP from ADP. Chemiosmosis is the process of the hydrogen ions flowing across the gated channel providing the energy for ATP synthase to work.**

BIOLOGY TEST #4 (Chapters 9-11)

1. How does DNA communicate with the rest of the cell? **Using mRNA and proteins. DNA contains the information that instructs the cell which proteins to make and when those proteins should be made. The message from the DNA is made into an RNA molecule. This message contains the instructions for the cell to make a specific polypeptide (protein). The RNA molecule then is read by a ribosome in the cytoplasm, and the protein is synthesized. The protein the cell makes causes the cell to behave in a certain way.**

2. What are the four nitrogenous bases of DNA? **Adenine (A), thymine (T), guanine (G), and cytosine (C).**

3. What is the main difference between the purines and pyrimidines, as far as molecular structure is concerned? **The purines have a double-ring structure in their nitrogen base; the pyrimidines have a single-ring structure.**

4. What does the term "complementary strands" mean? **This is often a hard concept for students to understand, and there may be several ways they can answer that are all correct. DNA exists as two complementary strands. One strand is complementary to the other because they form hydrogen bonds, which cause the two strands to stick together. Or, they are complementary because the sequence of one strand is such that every time there is an A in the strand, there is a T on the other strand directly across from it, so the A and T can base pair. Likewise, every time there is a C on one strand, there is a G on the other strand, and they base pair. The nucleotides on one strand always being able to base pair with the nucleotides on the other strand causes the two DNA strands to bond to one another and form the double-stranded molecule.**

5. What are the smaller units into which eukaryotic DNA is broken? **Chromosomes.**

6. What is a gene? **It is the individual functional unit of the chromosome. One gene contains the information for (or "codes for") the production of one protein.**

7. How is the cross shape of tRNA maintained? **Through hydrogen bonding between the nucleotides of the tRNA molecule.**

8. What is a codon? **It is a group of three nucleotides that code for the insertion of a specific amino acid into a specific location within a protein.**

9. Describe transcription and translation. Be sure to include the enzymes we discussed, mRNA processing, and signal sequences. **Transcription begins with helicase breaking the hydrogen bonds that hold complementary strands together, or unzipping the DNA. RNA polymerase binds to the DNA and begins to base pair RNA nucleotides using DNA as a template. When the gene is fully transcribed, mRNA is then processed, the introns are removed, and exons are spliced back together. mRNA leaves the nucleus through a nuclear pore and enters the cytoplasm. Then a ribosome binds the mRNA. The ribosome begins translating the mRNA at the start codon, AUG. Transcription proceeds with the ribosome, base pairing the appropriate tRNA anti-codon to the mRNA codon. While the tRNA is base paired to the mRNA, the ribosome catalyzes the condensation synthesis reaction, linking the amino acid that the tRNA holds to the growing protein (or polypeptide) chain. The ribosome proceeds along the mRNA molecule until it reaches a stop codon, then protein synthesis is completed. The protein then enters the endoplasmic reticulum and then the Golgi apparatus, where it is processed and has a signal sequence added to it. Following addition of the signal sequence, the protein is transported to wherever it is destined to go in the cell.**

10. True or False? Making mRNA from a DNA template ensures that the codons and reading frames are maintained. **True.**

11. How does the ribosome know that a particular tRNA molecule is carrying the proper amino acid the ribosome needs? **The tRNA has a section on it called the anticodon. The anticodon needs to be complimentary to the codon in order for the ribosome to know the tRNA is carrying the proper amino acid. If the anticodon is complimentary to the codon, they form hydrogen bonds (they base pair), which allows the ribosome to link the amino acid to the growing protein.**

12. Which bases base pair in DNA? **Adenine and thymine base pair, and cytosine and guanine base pair.**

13. True or False? Uracil base pairs with cytosine in RNA. **False. Uracil base pairs with adenine.**

14. What is cell division? **The process of one cell dividing into two cells.**

15. What is the theory of biogenesis? **The theory that life can only come from other living things.**

16. What type of reproduction is mitosis—sexual or asexual? **Asexual.**

17. Are cells formed from mitosis genetically identical or different? **Identical.**

18. What are the phases of mitosis? **Prophase, prometaphase, metaphase, anaphase, and telophase.**

19. When does DNA replication occur in eukaryotic asexual cell division? **At the end of interphase.**

20. If you are standing inside a cell and you see chromatin, is the cell in mitosis? **No, the cell is in interphase because the chromosomes are in the form of chromatin. When the cell is in mitosis, the DNA is visible as distinct chromosomes.**

21. If you are in a cell during mitosis and you see spindles forming from centrioles, are you in a plant or animal cell? Why? **You are in an animal cell because plant cells do not have centrioles.**

22. Describe the phases of mitosis and what is occurring to the chromosomes during each phase. Be sure to include cytokinesis and karyokinesis in your answer. **Prior to mitosis, at the end of interphase, the chromosomes are replicated. During prophase, chromosomes are condensing and are beginning to migrate toward the cell equator. The spindle is forming at this time, and the sister chromatids are attached to one another at the centromere. During prometaphase, the chromosomes have completely condensed and have nearly aligned completely at the equator. The spindle is formed. During metaphase, the chromosomes completely align along the cell equator. During anaphase, karyokinesis occurs. During telophase, the chromosomes begin to unwind, the nuclear membrane begins to form again, and cytokinesis occurs.**

23. True or False? Sister chromatids are identical copies of chromosomes attached to one another at the centromere. **True.**

24. How does the cell plate form and what happens to it? **The cell plate is found only in plants. It forms from the Golgi or ER and eventually develops into the cell wall between daughter cells.**

25. In sexually-reproducing organisms, what percent of the chromosomes come from the male parent and what percent from the female parent? **50% come from the male and 50% from the female.**

26. Describe the process of binary fission. **Binary fission starts with DNA replication. Each copy of the DNA attaches to the cell membrane at slightly different locations. The cell begins to elongate. As it does, the two DNA molecules move away from one another. Once the cell elongates enough, cytokinesis occurs (the cell membrane pinches inward), forming two genetically identical cells.**

27. Why are chromosomes paired? **Each chromosome of a pair contains genes that code for the same proteins (traits). Also, one chromosome of every pair comes from the male parent and one from the female parent.**

28. If an organism contains two X chromosomes, what sex is it? **Female.**

29. What is a diploid cell? **A cell that contains two copies of every chromosome (or a cell containing both pairs of every chromosome).**

30. What is a haploid cell? **A cell that contains one copy of every chromosome (or a cell containing only one chromosome of a pair).**

31. What is meiosis and where does it occur. **Meiosis is the biological process of making haploid gametes from diploid somatic cells. Meiosis occurs in the reproductive organs of the male and female organisms of every sexually-reproducing species.**

32. What is synapsis? **The process of the tetrads lining up at the equator of the cell.**

33. How many cells are formed after telophase I is completed? **Two.**

34. Are the cells formed after telophase I diploid or haploid? **Haploid.**

35. Following meiosis in the human male, how many sperm (gametes) are formed from one reproductive cell? **Four.**

36. True or False? When the male gamete fuses with the female gamete, the diploid number is restored. **True.**

BIOLOGY TEST #5 (Chapters 12-14)

1. What is heredity? **The passing of genetic information (traits) from generation to generation.**

2. What is the basic unit of heredity? **The gene.**

3. Describe what an allele is and how it is related to a gene and to the trait for which the gene codes. **An allele is an alternate form of a gene. A gene codes for a specific trait. There may be several different forms of the trait for which a gene codes. Each form of the trait is controlled by a different allele. For example, the trait of pea flower color had two forms—white and purple. There is a gene that controls the trait of flower color. The gene that codes for flower color has two alleles. One allele codes for a white flower, and the other allele codes for a purple flower.**

4. Which trait is expressed when an organism has a dominant and recessive allele of a gene and why? **The dominant trait is expressed because the dominant trait prevents the expression of the recessive trait.**

5. One allele, T, codes for long fingers and the other allele, t, codes for short fingers. What is the genotype and phenotype in this example? **The genotype is T and t. The phenotype is long fingers and short fingers.**

6. If an organism has the genotype Tt, what is the phenotype? **Since T is dominant to t (you know this because dominant alleles are always written with capital letters, and recessive alleles are always written with lower case letter); the phenotype encoded by t is suppressed by the phenotype encoded by T. Therefore, an organism with the genotype Tt will have the phenotype of long fingers.**

7. What will the genotype of the gametes be for an organism whose phenotype if Bb? **One of the gametes will be B and the other b.**

8. Can a Punnett square be used to predict expected genotype and phenotypes of offspring? **Yes. One of the major uses of Punnett squares is the ability to predict expected genotypes and phenotypes of offspring.**

9. What does independent assortment mean? **When genes segregate during meiosis, they separate independently. This means they are distributed to gametes without relationship to one another.**

10. True or False? An organism Bb for a condition is said to be homozygous recessive. **False. This organism is heterozygous; bb would be homozygous recessive.**

11. What is gene (allele) segregation and when does it occur? **It is the process of alleles that code for the same trait and located on opposite, paired chromosomes separating into different gametes during meiosis.**

12. If you are holding a six-sided dice, what is the probability that you will roll a 6? **The probability of that is equal to the outcome desired (rolling a 6) divided by the total possible outcomes (rolling a 1, 2, 3, 4, 5, or 6). This means there are six possible outcomes. The likelihood of rolling a 6 is 1/6. Other ways of stating that is a 1 in 6 chance or a 16.7 percent chance.**

13. What is the probability of rolling three successive 6s on a six-sided dice? **The probability of successive events occurring is equal to the product of the three independent events occurring. Since the probability of getting one 6 is 1/6, the probability of rolling three successive 6's is 1/6 x 1/6 x 1/6, or 1/216. Therefore the chance of this happening is 1/216, or 1 in 216 or 0.5%.**

14. Let's say the trait for seed color in sunflowers is controlled by one gene. There are four different alleles that code for this trait, each of which results in different seed colors. One of the four alleles—we will call it L—codes for black seeds. Another of the four alleles—R—codes for seeds that are black with white spots. Plants that are LL make black seeds. Plants that are RR make seeds that are black with white spots. Plants that are LR make both seeds that are black and seeds that are black with white spots. What type of inheritance pattern is this? **Co-dominance.**

15. Which chromosome carries most of the sex-linked traits? **The X chromosome.**

16. Is it impossible for a girl to have a recessive sex-linked trait? **No. It is unusual, but not impossible. As shown by the fruit fly experiments, a female can have a recessive sex-linked trait if she inherits a recessive sex-linked gene from her mom and the same recessive sex-linked gene from her dad.**

17. True or False? The inheritance pattern for fur color of an animal is incomplete dominance. There are two alleles—white and black. An organism that is heterozygous for fur color would have gray fur. **True.**

18. True or False? Epistasis describes a relationship between genes in which one gene that does not code for a trait modifies the expression of the gene that does. **True.**

19. Explain how environmental factors can influence the genotype of a pea plant that codes for a tall plant—for example, how the environment could "change" the phenotype. **If a pea plant that has a genotype coding for a tall plant is planted in poor soil conditions, the plant will grow short due to not having enough nutrients to grow tall.**

20. How do different alleles for the same gene arise? **Through mutations that alter the sequence of the gene in some way.**

21. What are the ways in which normal genetic variation can occur? **Independent assortment and crossing over.**

22. What is a change in the normal sequence of DNA called? **Mutation.**

23. True or False? Most of the space on DNA is taken up by genes. **False. Almost all the space of DNA is taken up by non-gene sequences. No more than 1% of DNA actually contains nucleotide sequences that code for the production of a protein.**

24. If a mutation occurs, is it more likely to occur in a gene or in a non-gene segment of DNA? **In a non-gene segment. Since the portion of DNA that does not code for a gene is so much larger than the portion that does, it is much more likely that a mutation will occur in a non-gene segment.**

25. What is a nonsense mutation? **It is a substitution mutation that changes an amino acid encoding codon to change into a stop codon.**

26. Why do deletions and additions alter the reading frame? **The reading frame is maintained by the orderly linear placement of nucleotides. When an addition or deletion mutation occurs, a nucleotide is either added or removed from its linear position. As a result, the reading frame changes. This is a difficult concept to put into words, but if the student can do it, they understand this material well. They could also use an example to show how the reading frame changes. I have several examples in the book they can use.**

27. Normal sentence: THE DOG RAN AND ATE THE PIG. Mutated sentence: THE DGR ANA NDA TET HEP IG. This is an example of a deletion mutation. The "O" in "DOG" was deleted. Why is the sentence so hard to read following the mutation? **The reading frame has shifted, and it does not make any sense.**

28. What is the basic problem of a nondisjunction mutation? **The chromosomes do not separate properly during karyokinesis. They "stick" together. This results in one gamete receiving two chromosomes of a pair and the other gamete receiving none.**

29. Why are germ cell mutations passed from parent to child? **Germ cell mutations occur in the gamete. This means the gamete is formed with a mutation present. All cells of an individual grow from the single cell formed when a male gamete fuses with a female gamete. If one of those gametes has a mutation, then the cell formed when that gamete fuses with another gamete will have the mutation. As the single cell undergoes mitosis to grow in size from a single cell to a multicellular organism, that mutation is passed to all the cells every time a cell divides. Then, when the organism is an adult (assuming it lives that long), half the gametes the organism produces will also have the mutation in them, so the mutation can be passed indefinitely.**

30. Substitutions, deletions, and addition are all types of _____mutations. **Point.**

31. True or False? Gene mutations involve larger segments of DNA than chromosomal mutations. **False.**

BIOLOGY TEST #6 (Chapters 15-16)

1. What is a congenital genetic disease? **A disease that a child is born with and is due to a mutation in the DNA.**

2. What is a carrier of a recessive condition? **A carrier is someone who has one normal/dominant allele and one recessive allele for a condition or disease.**

3. Does a carrier exhibit the recessive condition for which they carry the allele? **No. Since a carrier has one normal/dominant allele, its trait is expressed, so there is no effect of the recessive allele on a carrier.**

4. What does it mean when a genetic disease is said to be transmitted in an autosomal recessive fashion? **It means the allele for the disease is carried on one of the somatic chromosomes, or autosomes, and that the allele is a recessive allele, so the person needs to inherit two recessive alleles to have the disease.**

5. What is a pedigree analysis? **A pedigree analysis is a tool used to follow the passage of traits from one generation to the next. It shows the phenotypes of parents and their offspring from generation to generation. They are particularly useful in following the passage of recessive traits and diseases in families.**

6. What is heterozygous superiority? **It is the protection that the heterozygous condition can give to people who are carriers of a recessive allele.**

7. What is an autosomal dominant genetic disease? **It is a disease caused by a mutation on a somatic chromosome that is transmitted in a dominant fashion. Unlike most genetic diseases, it is caused by a dominant gene.**

8. Why are most people affected by X-linked (sex-linked) diseases male? **Almost all sex-linked diseases are transmitted on the X chromosome. Since males only have one X chromosome, they do not have the possibility of having a normal gene on their other sex chromosome. The Y chromosome does not carry many non-sex determining genes. Therefore, if a boy inherits am X chromosome from his mom that has a recessive gene on it for a disease, he will have the disease because there is not a normal, dominant gene to counteract it.**

9. True or False? Trisomy 21 is a type of aneuploidy. **True.**

10. Why is genetic engineering considered an area of great hope for treating diseases, especially congenital genetic diseases? **Many diseases may be able to be cured by giving the normal gene to a person with a disease caused by a defective gene.**

11. Describe how reverse transcriptase is used to make a molecule of double-stranded DNA. **Reverse transcriptase is added to the mRNA which codes for insulin, along with the nucleotides A, T, C, and G. After a while, a single-stranded DNA molecule is made. This single-stranded DNA molecule represents the gene for the insulin. The single-stranded DNA molecule is then mixed with DNA polymerase and more A, T, G, and C. After a while, the single-stranded DNA would be made by the DNA polymerase into a double-stranded DNA molecule.**

12. What are two important features of restriction enzymes? **They cut DNA at specific nucleotide sequences; when they cut the DNA, sticky ends are made.**

13. Describe the process of selective breeding if you were a farmer who wanted to breed fish to have large eyes. **The fish in question with the largest eyes would be collected and bred with one another. Over the generations, only the fish with the largest eyes would be allowed to breed with one another. Over time, the eyes would gradually become larger, as a result of the continual breeding of successive generations of large-eyed fish.**

14. Describe how one could possibly cure cystic fibrosis with gene therapy. **Since cystic fibrosis is a disease caused by a defective gene, being able to replace that defective gene with a normal gene could possibly cure this disease. When the normal gene is given to the pancreas and lung cells, it would allow these cells to start making the normal membrane channel protein, which pumps chloride.**

15. True or False? Dr. Griffith's experiments with *streptococcus* bacteria demonstrated that the traits of one organism cannot be altered by giving them DNA from another organism. **False. He proved the opposite.**

16. True or False? If you had restriction enzymes, a vector, and a host, you could make recombinant DNA. **True.**

17. What is polymerase chain reaction? **A technique to make large amounts of DNA.**

BIOLOGY TEST #7 (Chapter 17-19)

1. What is evolution? **Evolution is the scientific philosophy that all life on earth is here by accident; that over many billions of years, less complex organisms have acquired new traits, by chance, and changed into new organisms. Evolution is the belief system that all life on earth came from a single-celled bacterium.**

2. What is a creationist? **A person who believes that the origin of life is God's creation.**

3. What is the difference between an artifact and a fossil? **Fossils are the remains of a previously living life form. An artifact is past evidence specifically of human activity.**

4. Why are polystrate fossils a problem for evolutionists? **Because a polystrate fossil cannot have formed slowly.**

5. True or False? One of the reasons Darwin proposed the idea of evolution is because he was not a believer in God and it is hard to believe God created life if you do not believe in God. **True.**

6. True or False? Polystrate fossils, and the lack of organisms on the ocean bottom in various stages of fossilization, lend evidence to the argument that fossils do not form slowly, or at least *can* form rapidly. **True.**

7. True or False? Different radiometric dating techniques show excellent agreement in ages obtained for fossil specimens. **False. They often give dates that are nowhere near one another.**

8. If a person believes that all life forms are on earth due to an accidental arrangement of molecules that first formed single-cell organisms, then evolved directly into all life forms, what is the term used to describe that origin of life philosophy? **Evolution or naturalism.**

9. What is neo-Darwinism? **The modification of Darwin's theory to account for the contribution that genetics must have made for evolution to occur.**

10. True or False? Both creationists and evolutionists agree that natural selection occurs. **True.**

11. True or False? Both creationists and evolutionists agree natural selection and descent with modification occur, but evolutionists believe those processes can form new species, creationist do not. **True.**

12. Evolutionists state wings are a classic example of convergent evolution. What does this mean? How would creationists counter this statement? **Convergent evolution is when two or more unrelated organisms evolve into the same structure through descent with modification. God knew exactly what the needs of every organism He created was and would ever be. He created them with wings because He knew they would need them.**

13. According to most evolutionists and all creationists, does the fossil record contain intermediate forms? **No.**

14. True or False? Creationists believe God knew the needs of all organisms before He created them and gave them all the genetic variation they needed at creation. **True.**

15. What is the scientific problem that prevents neo-Darwinism from explaining the formation of new species? **Mutations do not add information, they only delete it. Since there is no known scientific way for information-adding mutations to occur, neo-Darwinism cannot possibly explain how a less-complex species evolves into a more-complex one.**

16. Why do creationists think gradualism is not a reasonable way to explain the origins of species? **Because if gradualism were true, there should be a fossil record full of transitional fossil forms. There are none. In addition, the appearance of species in the fossil record shows the abrupt appearance of fully-formed species. The species that are in the fossil record and still alive today have not changed at all. This goes against gradualism and evolution. Finally, there is at least one mass extinction in the fossil record. Evolutionists believe there are more, but that is based on the possibly-faulty logic that strata form slowly and fossils take thousands or millions of years to form. Evolutionists do not know what causes mass extinction. Creationists do—they say mass extinction was caused by the Great Flood of Noah.**

17. True or False? In order for neo-Darwinism to explain evolution, unproven gene-adding mutations need to occur. **True.**

18. If gene-adding mutations do not explain how organisms develop new traits and transform into new species, can neo-Darwinism be true? **No.**

BIOLOGY TEST #8 (Chapters 20-21)

1. If an organism's binomial name is *Bacillus anthrasis* what is its genus? **Bacillus**. What is its species? **Anthrasis**.

2. If you are studying prokaryotic species, into which kingdom(s) could they be classified? **All prokaryotes are classified into Archaebacteria or Eubacteria.**

3. What are the defining features of organisms from Protista? **They are unicellular and eukaryotic.**

4. What are the defining characteristics of the organisms from Fungi? **They all have cell walls made of chitin and are decomposers.**

5. What are the defining characteristics of Plantae? **They all have cell walls made of cellulose, perform photosynthesis, and are multicellular.**

6. What are the defining characteristics of the organisms of Animalia? **Multicellularity, heterotrophic, no cell walls, sexual reproduction, mobility, eukaryotic, storage of excess energy as glycogen, cells organized into tissues, cells held together by extra-cellular matrix, aerobic, and have a symmetric body plan.**

7. Why are viruses not considered "alive"? **Viruses do not exhibit several properties that all true forms of life exhibit. Viruses do not grow, do not independently reproduce, do not perform any of their own metabolic processes, and cannot maintain homeostasis.**

8. Why are retroviruses important in recombinant DNA? **Retroviruses contain reverse transcriptase, a critical enzyme in some recombinant technology.**

9. What does a prion contain that makes it infectious? **A protein.**

10. True or False? Aristotle invented the modern classification system more than two thousand years ago. **False. Linnaeus invented it in the 1700s.**

11. What are the three common bacterial shapes? **Rod (bacillus), sphere (cocci), and spiral (spirochete).**

12. True or False? All bacteria produce endospores. **False. Some bacteria form endospores, some do not.**

13. What molecules make up Eubacterial cell walls? **Peptidoglycan.**

14. What is a capsule? **A polysaccharide coating that some bacterial species have.**

15. What are endotoxins and which type of bacteria produces them? **It is a mixture of lipids and carbohydrates formed from the outer membrane when a Gram-negative bacterium breaks apart.**

16. What are exotoxins and what type of bacteria produces them? **They are proteins secreted by some Gram-positive bacteria.**

17. How do antibiotics such as penicillin and cephalosporin work? **They inhibit a bacteria's ability to make the cell wall.**

18. What is an obligate anaerobe? **They are bacteria that cannot live in the presence of oxygen.**

19. Why are the organisms of Eubacteria and Archaebacteria separated into two kingdoms? **Because the cell walls of the Eubacteria and Archaebacteria are structurally different. Also, the sequence of bases in the tRNA and rRNA of the two kingdoms is different.**

20. True or False? A Gram-negative bacteria stains pink during Gram staining because they have thicker cell walls than Gram-positive bacteria. **False. Gram-negative cells have a thinner cell wall than Gram-positive cells.**

BIOLOGY TEST #9 (Chapters 22-23)

1. How are the protists often broken down? **The animal-like protists, or protozoa. They are also called heterotrophic protists. The plant-like protists, or algae. They are also called the autotrophic protists. The fungus-like protists, or slime molds and water molds. They are also called the absorptive protists.**

2. What is a stipe? **It is the stem-like part of a multicellular alga.**

3. What is alternation of generations? **It is the description of the life cycle in which the organism exists in the haploid state in one generation and in the diploid states in the following generation.**

4. What is a pseudopod, and how does it form? **It is an extension of the cell membrane that forms as a result of cytoplasmic streaming.**

5. True or False? Phytoplankton is the protozoan component of plankton. **False. It is the algal component.**

6. True or False? Most algae are unicellular. **True.**

7. What is the vector and host for *Plasmodium*? **The vector is the *Anopheles* mosquito and the host is the human.**

8. How do Fungi obtain their nutrition? **Fungi are saprophytes. They perform extracellular digestion, then absorb the nutrients.**

9. What is a non-septated hyphus called? **A coenocytic hyphus.**

10. What is a cap? **It is the top of a basidiocarp (mushroom) that contains the reproductive structures.**

11. What are rhizoids and stolons? **Rhizoids are hyphae that grow into the surface of whatever a fungus grows on. Stolons are hyphae that grow over the surface of whatever the fungus is living on.**

12. True or False? All fungi have cell walls made of chitin. **True.**

13. What is lichen? **A symbiotic relationship between a fungal species and an algal or photosynthetic bacterial species.**

14. True or False? Basidiocarps, Ascocarps, and rhizoids are all structures that can be found in members of Protista. **False. They are found in members of Fungi.**

BIOLOGY TEST #10 (Chapters 24-25)

1. True or False? Not all plants are autotrophic. **False. One of the defining features of plants is that they are all photosynthetic, which means they are all autotrophic.**

2. What is one organelle that plant cells do not have but animal cells do? **Centrioles.**

3. What is the primary cell wall? **The first cell wall made by a plant cell.**

4. What is the middle lamella made of and what is its function? **It is made of pectins, which are sticky and hold together adjacent primary cell walls.**

5. Which cell wall is stronger, the primary or the secondary? **The secondary.**

6. How are a seed and a spore different? **A spore contains only one cell with the ability to grow into a new plant. Spores also do not have any type of protective coating, as a seed does. A seed is a structure with a protective coating that contains more in its inside than a single cell. A seed contains a plant embryo surrounded by a protective coating. An embryo is a multicellular "mini plant," which will grow into a new plant. A seed also contains endosperm. Seeds are also much larger than spores.**

7. What is the difference between a monocot and a dicot seed? **A monocot seed has one cotyledon, or seed leaf, and a dicot has two.**

8. Where does plant growth take place? **Meristem tissue.**

9. What is the difference in xylem and phloem location between monocot and dicot roots? **In monocot roots, the xylem and phloem are organized like spokes on a wheel. In dicot roots, the xylem is oriented like an X, and the phloem is in between the arms of the X.**

10. What is the difference is location of the xylem and phloem in a monocot and dicot stem? **A monocot stem has the vascular bundles scattered throughout the stem. Dicot stems have the xylem and phloem form in rings around each other.**

11. What vascular tissue is located in the inner bark layer? **Phloem.**

12. How do tree rings form? **In the springtime, when water is plentiful, a lot of xylem is needed to transport it, so more xylem is made in the spring than in the late summer, fall, and winter. Therefore, the spring xylem is larger and not as densely packed as the later xylem. When a tree is cut down, the spring xylem is lighter in color; the later xylem is darker in color. During the winter, when no xylem is made, a dark line forms between the late xylem of one season and the spring xylem of the new growing season. The number of rings present indicates how old the tree is, and the thickness of the rings gives an idea of how harsh the climate was for any particular growing season (year).**

13. What is responsible for nastic movements in plants? **Turgor pressure.**

14. Describe how the cohesion-tension theory explains the movement of nutrients throughout the plant. **The cohesion-tension theory starts with water loss in the leaves of the plant. During the daytime, when water is plentiful, the guard cells keep the stoma open, so the leaves can exchange gas and maximally perform photosynthesis. Photosynthesis causes the plant to lose water. Also, there is evaporation of the water directly from the leaves, called transpiration. Both photosynthesis and transpiration lead to significant water loss in the plant occurring at the leaves, meaning there is an overall deficit of water in the leaves. As water exits the plant from the leaves, the remaining water in the xylem of the branches, trunk, and stem want to stick to the water that is being lost in the leaves. This causes tension to develop because the amount of water in the leaves is less than the amount in the rest of the plant. As the water exits the leaves, the rest of the water in the plant wants to stick to that water. This tension "pulls" the rest of the water molecules up toward the leaves from further down the plant. Since the nutrients absorbed in the roots are dissolved in the water, they go along for the ride, and there is a continual transport of water and nutrients up the plant from the roots.**

15. True or False? Xylem carries material from leaves to the rest of the plant, and phloem carries material from the roots to the rest of the plant. **False. Xylem carries from the roots and phloem from the leaves.**

16. True or False? Nonvascular plants grow much larger than vascular plants. **False.**

17. What plant hormone is responsible for most tropisms? **Auxin.**

18. When you look at a vascular plant, are you looking at a gametophyte or sporophyte? **A sporophyte. In the vascular plants, the sporophyte stage is dominant, and the gametophyte is usually so small that it is not easy to see.**

19. In which structures do mosses produce the male and female gametes? **The male gametes are formed in the antheridium and the female gametes are formed in the archegonium.**

20. What are the parts of the pistil? **The ovary, ovules, style, and stigma.**

21. True or False? Gametophytes are gamete-producing plants. **True.**

22. True or False? Anthers, filaments, and stamens are all parts of male plant reproductive structures. **True.**

23. Describe the process of angiosperm pollination and fertilization. **Pollination is the process of pollen, the male gamete, getting transferred from the anther to the stigma. This can occur through the action of the wind or through pollinators, such as insects or small animals. Once the pollen lands on the stigma, the generative cell inside divides into two sperm cells. The tube cell inside the pollen grows down the style to the micropyle of the ovary. Then both sperm proceed down the tube to the ovary. One sperm fertilize the egg; the other sperm fertilizes two polar bodies. The egg and sperm form the embryo, and the sperm and two polar bodies form the endosperm. A tough coat, called a seed coat, forms around the endosperm and embryo, forming a seed.**

24. True or False? Sporophytes are diploid. **True.**

25. True or False? Gametophytes produce haploid spores through meiosis. **False. Gametophytes produce haploid gametes by mitosis.**

26. True or False? Angiosperms display "double fertilization." **True.**

BIOLOGY TEST #11 (Chapter 26-27)

1. What is a vertebrate? **An animal with a backbone, spinal column, or vertebrae.**

2. What is cephalization? **The property animals display when the head end becomes larger due to the concentration of sensory structures in the head.**

3. What is a zygote? **It is the cell formed when a sperm fuses with an egg.**

4. What happens to germ layers as the gastrula grows in size? **They differentiate into different types of tissue.**

5. How does a sponge feed? **Sponges are filter feeders. They use flagella to move water and nutrients through openings, called pores, in the body. Sponge cells extract nutrients directly from the water.**

6. True or False? In vertebrates, the notochord develops into the spinal cord, and the neural tube develops into vertebrae (of the spinal column). **False. The notochord develops into the vertebrae, and the neural tube into the spine.**

7. What is a distinguishing feature of Cnidaria? **The presence of special stinging cells called cnidocytes.**

8. What is a scolex? **It is the specialized head segment of a tapeworm, which is specialized to attach to the host.**

9. True or False? Nemertines are the simplest phylum to exhibit a fully-formed digestive tract and a circulatory system. **True.**

10. What is an exoskeleton? **It is a hard protective covering on the outside of the organism.**

11. What is the defining feature of animals classified in Chordata? **The presence of a notochord at some point in their life cycle.**

12. True or False? There are many times more vertebrates than invertebrates. **False.**

13. _____symmetry is when the right half of an organism looks like the left half.

14. True or False? Most animals develop from two germ layers. **False. Most animals develop from three germ layers.**

15. True or False? The body of all mollusks is soft. **True. Although clams have hard shells, their inner body is soft.**

16. True or False? A head-foot is part of the body plan of Annelida. **False. It describes the body plan of mollusks.**

17. Why do insects molt? **Since an exoskeleton is rigid, it does not allow for the organism to grow. As an organism grows with an exoskeleton, the exoskeleton peels off at various times to allow for the growth of the organism. This process is called molting. The main key is that the student understands what molting is and that it happens as a result of the exoskeleton not allowing growth to occur.**

18. Describe the process of complete metamorphosis of a caterpillar to a butterfly. Be sure to include technical terms for all stages. **The caterpillar is the larval stage. It spins a cocoon (called a chrysalis) and enters the next stage (called a pupa). Inside the cocoon, the pupa transforms into a butterfly. The butterfly is the adult stage.**

19. True or False? A nymph is an immature form of an organism from Chordata. **False. Chordates don't have larval or nymph stages.**

BIOLOGY TEST #12 (Chapter 28-29)

1. What is the defining feature of organisms classified in Vertebrata? **The presence of a dorsal spinal cord protected or surrounded by a spinal column.**

2. What is a defining feature of Chondrichthyes, like sharks? **Their skeletons are made out of cartilage.**

3. How many chambers do fish hearts have? **Two.**

4. Describe the circulatory pathway in a fish, starting in the atrium. Is it a one-loop or two-loop system? **The circulatory system of fish is a one-loop system. Starting in the atrium, the blood flows from the atrium to the ventricle, then to the gills, then to the tissues. From the tissues, it flows back to the atrium.**

5. What are the two loops called of a two-loop circulatory system? **One loop is the pulmonary loop, or pulmonary circulation; the other is the systemic loop, or systemic circulation.**

6. Why are reptiles less dependent on water than amphibians? **They have watertight skin, so they do not need to keep their skin wet all the time, as amphibians do. Also, they produce amniotic eggs, which protect the developing embryo from drying out. Amphibians must lay their unprotected eggs in or close to water.**

7. What is the difference between an endotherm and an ectotherm? **Endotherms generate enough heat through their metabolic processes to maintain temperature homeostasis, regardless of the environmental temperature. Ectotherms do not have as high of a metabolic rate and must rely on the environment to regulate their body temperature.**

8. True or False? Amphibians are dependent on water because their eggs need to stay moist, as does their skin. **True.**

9. True or False? Birds, reptiles, and some mammals lay amniotic eggs. **True.**

10. True or False? Some of the organisms classified in Aves are not endotherms. **False. All birds are endotherms.**

11. What is the general consensus regarding *Archaeopteryx*? **It is a bird fossil. Almost all the leading ornithologists who have studied it think it is a bird.**

12. Describe the special way birds breathe. **Air enters into the bird through the nostrils and goes not only into the lungs, but also fills the air sacs. During inspiration (breathing in), the lungs extract oxygen and release carbon dioxide while the air sacs are filled with air (no gas exchange occurs in the air sacs as they simply store the air during inspiration). During exhalation (breathing out), the air in the lungs—now containing carbon dioxide—rushes out of the lungs and is replaced with the air from the air sacs, which still has oxygen in it. This allows the lungs of birds to continually exchange gas as the lungs are constantly filled with oxygen-containing air. In addition, the air sacs help significantly reduce the weight of the bird.**

13. What are the three general categories of Mammalia? **Monotremata, "the egg-laying mammals"; Marsupialia, "the pouched mammals"; Placentalia, "the placental mammals."**

14. List at least five features birds have in common. **Beaks, wings, feathers, amniotic eggs, lightweight skeletons, keel, endothermia, oviparity, air sacs. NOT flight.**

15. True or False? Like all aquatic animals, aquatic mammals have gills. **False. All mammals have lungs.**

16. True or False? Marsupials give birth to immature, live young, but do not form placentas. **False. Marsupials do form placentas.**

17. List at least five characteristics all mammals have in common. **Hair, mammary glands, single jaw bone, high level of parental care, large brains, endothermia, lungs, specialized teeth, store fat and glycogen.**

18. True or False? Sirens and Cetaceans are able to use echolocation. **False. Only cetaceans and chiroptera can.**

BIOLOGY TEST #13 (Chapters 30-31)

1. What is the basic purpose of the nervous system? **It receives information from the environment and processes it to react in some way to the information presented to it.**

2. What is the Na-K-ATP pump, and how does it maintain the resting potential? **The Na-K-ATP pump actively pumps sodium out of the neuron and potassium into the neuron. This maintains a resting potential, or charge, of -70mV on the inside of the nerve cell as compared to the outside of the cell.**

3. What happens if a stimulus does not reach threshold for a neuron? **Nothing.**

4. What is contained in the gray matter of the brain? **The cell bodies of brain axons.**

5. Which part of the ANS gets you ready for fight or flight? **The sympathetic nervous system.**

6. Describe the mechanism of hearing. **Sound waves pass into the auditory canal and strike the ear drum (tympanum). The ear drum is connected to the three ear bones (ossicles), and when the sound hits the ear drum, it vibrates. When the ear drum vibrates, it causes the ossicles to vibrate. The ossicles transmit the vibration into the cochlea. The cochlea is filled with fluid, and when the ossicles vibrate, the fluid in the cochlea moves. The movement of the fluid causes movement of little hairs on the ends of nerve cells. As the hairs move, nerve impulses are generated that are transmitted through the acoustic nerve to the brain and processed.**

7. True or False? Anatomy is the structure and physiology is the function of living things. **True.**

8. What part of a neuron brings information into the cell body? **The dendrite.**

9. True or False? Axons are sometimes covered in layers of fat called myelin. **True**.

10. True or False? When a nerve depolarizes, the Na-K-ATPase pump temporarily turns on. **False. It temporarily shuts off.**

11. Describe how a nerve repolarizes. **The Na-K-ATPase pump turns on and pumps Na+ out of the cell and K+ into the cell.**

12. True or False? Saltatory conduction is possible because of myelin. **True.**

13. True or False? A neurotransmitter is a chemical nerves use to communicate with one another. **True.**

14. True or False? The brain stem fine-tunes and controls movements. **False. The cerebellum does.**

15. Which part of the brain is responsible for processing vision? **Occipital lobes.**

16. Describe the pathway a reflex takes. **The stimulus (or sensation) travels up the sensory nerves (dendrites) to the spinal cord. An interneuron in the spinal cord transmits the impulse from the sensory nerve to the motor nerve in the spinal cord. Then the motor neuron axon carries the signal to move to the muscles.**

17. True or False? Cones are responsible for color vision. **True.**

18. True or False? The taste sensations are scattered randomly over the tongue. **False. They are located in specific areas.**

19. What is a pathogen? **An infectious organism; something that can cause disease.**

20. What are some specific examples of chemical barriers to prevent pathogens from entering the body? **They are found mainly in the fluids of the eyes, mouth, and stomach. These chemicals are usually enzymes that function to break down the cell wall or membrane of the pathogen. They are present all the time, on standby in case a pathogen gets past the physical barriers. In addition, the acidic environment of the stomach often kills pathogens.**

21. Describe how an immune response is activated. **A foreign invader escapes the barriers and the nonspecific response. Macrophages engulf and process the invader and present antigens from the invader to the t-helper cell. The t-helper cell takes the antigen to the b-cells and cytotoxic t-cells. This allows the b and t-cells to identify the invader and prepare to specifically attack it. The b-cell makes antibodies, which will be secreted into the blood and stick to the antigen on the pathogen. The cytotoxic t-cell will make special molecules on its cell surface, called receptors, which will stick to the antigen on the pathogen. Many b-cells and cytotoxic t-cells are cloned. These all attack only the pathogen with the antigen on it. The antibodies make it easy for neutrophils and macrophages to target and engulf the pathogens. The cytotoxic t-cells stick to the pathogens and kill them. In this way, the pathogen is specifically targeted for removal.**

22. What type of immunity is acquired as a result of immunizations? **Active.**

23. What is a hormone? **Hormones are the chemical messengers of the endocrine system.**

24. What is a target organ and what do they have for hormones? **It is the organ with which the hormone is meant to communicate. Every target organ has a receptor on it, which sticks to the hormone. When the hormone passes by the target organ, it binds to the receptor.**

25. True of False? Amino acid hormones' receptors are found on the cell surface because amino acid hormones are not soluble in the cell membrane. **True**

26. If the thyroid is producing too much hormone, does the hypothalamus produce TRH (thyrotropic releasing hormone)? **No, because when TRH is released, it causes TSH to be released from the pituitary. TSH causes more thyroid hormones to be released from the thyroid. Therefore, TRH production would decrease.**

27. What is a nephron and what is it composed of? **A nephron is a functional unit of the kidney. It is composed of a glomerulus and Bowman's Capsule.**

28. If you sampled and tested a urine specimen and found it had high levels of protein and glucose, would you conclude that person's kidneys were sick or healthy? **Sick.**

29. Match the systems with its correct function:

nervous	**receives and processes information from the environment and effects actions**
endocrine	**chemical control of body functions**
musculoskeletal	**structure, support, movement**
immune	**protection from infection**
digestive	**obtain, break down, and absorb nutrients**
excretory	**rid body of wastes, but keep nutrients**
circulatory	**move nutrients, gasses, & molecules around the body**
respiratory	**bring air in and out of the body for gas exchange**
integumentary	**cover and protect body**
connective	**hold body together**

BIOLOGY TEST #14 (Chapters 32-33)

1. What are the components and functions of the circulatory system? **The heart pumps the blood through the blood vessels. Arteries carry blood away from the heart. Veins carry blood toward the heart. Capillaries are very small vessels where gas and nutrient exchange occurs. Blood is technically not part of the circulatory system, but its contribution to the circulatory system cannot be ignored.**

2. What are the two large, muscular chambers of the heart called? **Ventricles.**

3. Is the blood in the pulmonary vein oxygenated or deoxygenated? **It is oxygenated because it has gone through the lungs and released carbon dioxide and picked up oxygen.**

4. Describe the pathway of the electrical impulse for the heartbeat. **The signal for the heartbeat starts in the sinoatrial (S-A) node. It is then conducted through the atria, and the atria contract. The signal is then delayed slightly at the atrioventricular (A-V) node. After a short delay, the signal is conducted into the ventricles, and they contract.**

5. What are the components of blood? **Red blood cells, white blood cells, platelets, and plasma.**

6. What is an alveolus? **It is an air sac in the lung where gas exchange occurs; it is the functional unit of the lung.**

7. What is ventilation? **It is the process of moving air in and out of the lungs.**

8. What is respiration? **It is the process of gas exchange.**

9. Arrange these structures in order of the route blood takes, starting from the left atrium. You can number them in their order, or re-write them in the proper order using arrows to indicate the direction of the blood flow.

left atrium	**left atrium**
pulmonic valve	**mitral valve**
systemic arteries	**left ventricle**
mitral valve	**aortic valve**
right ventricle	**aorta**
systemic capillaries	**systemic arteries**
left ventricle	**capillaries**
systemic veins	**systemic veins**
aortic valve	**superior or inferior vena cava**
pulmonary arteries	**right atrium**
tricuspid valve	**tricuspid valve**
right atrium	**right ventricle**
aorta	**pulmonic valve**
superior or inferior vena cava	**pulmonary arteries**
pulmonary capillaries	**pulmonary capillaries**
pulmonary veins	**pulmonary veins**
	left atrium

10. True or False? The mitral valve closes when the left ventricle contracts. **True.**

11. True or False? The pulmonid valve closes when the right ventricle contracts. **False, it opens.**

12. What is hemoglobin, where is it found, and why is it important. **It is a large protein found in red blood cells. It carries oxygen to our tissues.**

13. Describe the process/mechanics of how we inhale and exhale. **The diaphragm contracts and moves downward in the abdomen. This increases the chest cavity space and creates negative pressure in the chest. The negative pressure in the chest causes the lungs to expand, which creates negative pressure inside the lungs and a vacuum-like effect, causing air outside of the body to be drawn into the respiratory tract. This fills the lungs with air. Exhalation occurs because the diaphragm relaxes, and the elastic recoil of the lungs pulls them into a smaller size, forcing air out of the respiratory track/lungs.**

14. What do teeth do? **They mechanically break food into smaller pieces.**

15. What is peristalsis? **Waves of muscular contractions that propel food through the GI tract.**

16. What is chyme? **It is the acidic paste mixture of food after it is churned and broken down by the stomach.**

17. What do the pancreas and liver empty into the duodenum, and how do these chemical contribute to digestion? **The pancreas releases many enzymes into the duodenum to further break down fats, proteins, and carbohydrates. The liver makes bile and stores it in the gall bladder. The bile is released into the duodenum and helps the enzymes of the pancreas break down fats better.**

18. What are villi and microvilli and why are they important? **Villi are foldings of the inside of the small intestine and microvilli are foldings of the villi. They are important because the foldings greatly increase the absorptive ability of the small intestines. All the nutrients, vitamins, and minerals needed are absorbed by the villi and microvilli.**

19. What is the function of ligaments and tendons? **Ligaments hold bones together (connect bone to bone) and ligaments hold muscles to bones.**

20. What is the periosteum? **It is the membrane covering all bones.**

21. What are Haversian canals? **They are canals in the compact bone that allow blood vessels and nerves to pass.**

22. What is a ball-and-socket joint? Give an example? **It is a joint that allows movement in all planes. Examples are the hip and shoulder joints.**

23. What are the differences between smooth and skeletal muscle? **There are several answers for this question. Smooth muscle is not under voluntary control; skeletal muscle is. Smooth muscle does not have striations; skeletal muscle does. Smooth muscle cells contain one nucleus per cell; skeletal muscle cells contain multiple nuclei per cell.**

24. What is acetylcholine, and what happens when it is released at the neuromuscular junction? **Acetylcholine is the neuromuscular transmitter. When the nerve releases it, it causes the muscle to contract.**

25. What are the cells called that make up the skin? **Epithelial cells.**

26. What is the lower layer of the skin called? **The dermis.**

27. What are the organs or structures that mechanically break food down? **Teeth and stomach.**

28. What are the organs or structures that chemically break food down? **Salivary glands, stomach, pancreas, and liver/gall bladder.**

29. True or False? Pepsinogen is an inactive form of the enzyme pepsin. **True.**

30. True or False? Lacteals are responsible for absorbing proteins. **False. Lacteals absorb fats.**

31. True or False? Compact bone is found surrounding spongy bone. **True**

32. True or False? Most bones form through the process of chondrification. **False. They form through ossification.**

33. True or False? Hinge joints allow for greater movement than ball-and-socket joints. **False. Ball and socket joints provide the greatest movement.**

34. True or False? A sarcomere is the functional unit of a muscle cell. **True.**

35. True or False? Actin and dystrophin are contractile proteins that make the sarcomere contract (get shorter). **False. Actin and myosin are the contractile proteins that make the sarcomere contract.**

36. True or False? Follicles and oil glands originate in the epidermis. **False. They originate in the dermis.**

37. True or False? The dermis is composed of a layer of living and dead cells. **False. The dermis is all living cells. The epidermis is part living and part dead cell layers.**

BIOLOGY TEST #15 (Chapter 34)

1. What is the study of the relationships living organisms have with one another and their physical environment called? **Ecology.**

2. What is an ecosystem? **The association and interaction of all living organisms within their physical environment.**

3. Which would you usually expect to contain a larger number of organisms in an ecosystem, a population or community? **Normally a community would be larger because a community includes all the organisms in an ecosystem and a population includes all the members of a certain species in an ecosystem.**

4. If you are studying an ecosystem and find that one of the populations has a negative growth rate, what does that mean? **The population is getting smaller. A negative growth rate indicates the population is not getting larger, but smaller.**

5. What is an accurate transfer of energy between all organisms in an ecosystem called? **A food web.**

6. The organism that is on the bottom of the food pyramid has the highest or lowest amount of biomass and energy available to the ecosystem? **The highest.**

7. How does transpiration differ from evaporation? **Evaporation is water on the earth's surface turning into water vapor. Transpiration is the loss of water by plants into the atmosphere.**

8. What is a watershed? **It is an area where runoff water accumulates.**

9. How is most of the atmospheric oxygen replenished? **Photosynthesis.**

10. How is atmospheric carbon dioxide replenished so it can be used for photosynthesis? **Cellular respiration, burning fossil fuels, and dead organism decomposition.**

11. Is the greenhouse effect "bad." **No. Without the greenhouse effect there would be no life on earth because the earth would be too cold.**

12. What is a biome? **A biome is a major regional group of distinctive plant and animal species in a common physical environment.**

13. What is a photic zone? **A photic zone is an area of an aquatic biome where light can penetrate and photosynthesis can occur.**

14. What is the difference between biotic and abiotic mass? **Biotic mass is the living mass/ organisms in an ecosystem; abiotic mass is the non-living mass.**

15. True or False? A clumped population distribution pattern is characterized by random distribution of organisms in an ecosystem. **False. It is characterized by groups of organisms.**

16. True or False? A primary consumer is one that eats a producer. **True.**

17. True or False? Mimicry exists when an organism is shaped or looks like part of its surroundings so it blends in and is not seen easily. **False. That is the description of camouflage. Mimicry exists when a harmless specie looks like a harmful one.**

18. True or False? Biochemical cycles describe the flow and recycling of atoms, elements, or molecules in an ecosystem. **True.**

19. True or False? A tundra biome has a lower average temperature than a desert. **True.**

20. True or False? Population density gives an idea about how crowded the population is. **True.**

21. True or False? Carbon dioxide is a greenhouse gas. **True.**

22. How are flora and fauna different? **Flora refers to plants in an ecosystem, fauna refers to animals living in an ecosystem.**

23. What is a pioneer specie? **The first organism to populate a new ecosystem.**

24. What is a climax community? **A stable population of flora and fauna in an ecosystem.**

PARENT COMPANION

1 | Introduction

1.0 CHAPTER PREVIEW

In this chapter we will:
- Discuss the following properties that all living things have in common:
    - They are made up of one or more cells.
    - They contain DNA.
    - They reproduce.
    - They are complex and organized.
    - They respond to their surroundings.
    - They extract energy from their environment.
    - They maintain homeostasis.
    - They grow.
    - They can be systematically classified.

- Explore the principles and practice of the scientific method.

- Discuss the SI measurement units.

- Discuss properties of light and electron microscopes.

Topic question

None

Christian perspective

God gave life to every living thing on earth.

1.1 INTRODUCTION TO BIOLOGY

- Biology is the study of life.
- Although it is often easy to tell what is and what is not alive, there are properties of life that all living things share.

Topic question

Is a tree alive? **Yes.** Is a rock alive? **No.** Is a starfish alive? **Yes. A starfish is actually an animal. (This is sometimes hard for students to know but we will learn this later in the year, so do not worry if they don't know it right now).**

Is the wood you burn in the fireplace alive? **No. Although it once was alive, as soon as it is cut down, the wood from the tree is dead.**

1.2 CELLS

- The cell is the basic functional unit of life.
- Some organisms consist of one cell, or are unicellular.
- Other organisms are more than one cell, or are multicellular.

Topic question

There are organisms that can be seen with the eye and ones that cannot be seen with the eye. Does it make sense that the organism which can be seen with the eye are multicellular? **Yes. Since they are multicellular, they consist of more than one cell. The more cells an organism is made up of, the larger it will be. This is a question that will allow the student to think outside the box and use their intuition.**

1.3 DEOXYRIBONUCLEIC ACID (DNA)

- All living things contain the molecule DNA (deoxyribonucleic acid).

- DNA contains all the information that a cell and organism need to function properly.

- DNA also is the reason that all organisms look and act the way they do. It is what makes all organisms unique.

Topic question

Your student may have heard of something called a virus. As they will learn, viruses are not living for a number of reasons. However, tell your student that there is such an organism as a virus, but this virus only contains a molecule called RNA. It does not contain DNA. Is this organism "living"? **The answer is no because one of the criteria for being alive is that an organism contains DNA. If it contains no DNA, then it is not alive. Again, I am trying to encourage them to think outside of the box and apply what they have learned to a situation they have not directly learned about.**

Christian Perspective

God is the ultimate scientist. He devised this extremely complicated molecule called DNA. Within the DNA is all the information needed for organisms to survive.

1.4 REPRODUCTION

- All life forms reproduce.

- There are two kinds of reproduction: sexual and asexual.

- Asexual reproduction results in the formation of two identical organisms from one organism.

- Sexual reproduction requires a male and a female organism to accomplish. It results in the combination of DNA from the male and the female organisms, which results in an organism completely different from the male or the female.

- Traits are passed from the parent organism to the offspring during sexual and asexual reproduction.

Topic question

Why is asexual reproduction similar to making photocopies? **Because both asexual reproduction and making photocopies result in the production of identical objects. In the case of asexual reproduction, the offspring is identical to the parent. In the case of the photocopy, the copy is identical to the original.**

Christian perspective

God made all organisms reproduce after their kind. This means that giraffes produce only giraffes and maple trees produce only maple trees. This becomes important in the discussion of evolution later in the year.

1.5 COMPLEX AND ORGANIZED

- All life forms are highly complex and organized.

- This organization and complexity are seen on all levels of life.

Topic question

Although individual organisms are complex and organized, their interactions with one another and their surroundings are not. True or False? **False. Everything about organisms/life is complex and organized.**

1.6 RESPONSIVE

- All living things respond to their surroundings.

Topic question

Why is a rock or a light pole not alive? **One of the most common reasons students give is because they cannot move. I want the student to realize that means nonliving things cannot respond to their environments. However, if they say because they do not exhibit any of the other properties of life, that is a fine answer, too.**

1.7 ENERGY EXTRACTION

- All organisms obtain energy for their life processes from their environment.
- Plants make their own energy through photosynthesis. Plants are therefore called producers.
- Animals that eat plants and other animals to obtain their energy are called consumers.
- The food chain is the relationship which consumers have to producers and to one another.

Topic questions

If an animal eats another animal to obtain its energy, what is that animal called? **A consumer, a carnivore or a predator are all OK answers.**

If an organism eats the dead remains of another organism to obtain its energy, what is that organism called? **A decomposer.**

1.8 HOMEOSTASIS

- All living organisms maintain stable internal living environments. This is called homeostasis.

Topic question

What is homeostasis? **The maintenance of a stable internal environment.**

1.9 GROWTH

- All living organisms grow.

Topic question

Since adult organisms no longer grow, why are they considered alive? **They are still growing even if you cannot "see" that they are. All life forms do grow regardless of their age. The growth may be on a microscopic level, but they still grow. For example, adult humans are continually growing new hair and new skin cells. Just because an organism does not grow taller or larger does not mean they do not grow!**

1.10 CLASSIFICATION

- All living things are classified based on common characteristics. The study of classifying living organisms is called taxonomy.
- There are seven levels of classification used in modern biology.

Topic questions

What are the seven levels of classification? **Kingdom, Phylum (or Division for plants and fungi), Class, Order, Family, Genus, and Species.**

What is the least specific level of classification? **Kingdom**.

What is a binomial name? **It is the most specific level of classification for an organism. It is the Genus of the organism followed by the species. All binomial names are Latin.**

1.11 SCIENTIFIC METHOD

- The scientific method is a methodical way to obtain and interpret data.
- The steps of the scientific method are: hypothesis statement, experimentation, and data analysis.
- There are different ways to conduct scientific experiments.

Topic question

In an experiment, describe the difference between the experimental and the control groups. **The experimental group is exposed to the exact same conditions or factors as the control group except one. The factor that is different in the experimental group is called the variable. This type of experiment is designed to evaluate the effects of the variable on the organism.**

1.12 THEORY

- A theory is a hypothesis that has withstood rigorous testing over time and been found to be true.

Topic question

If new information is uncovered as a result of further experimentation that does not support a previous "theory," what must one do? **This situation occurs in science. Something which was previously thought to be "true," or was considered a theory, is proven not to be true. It is required that all scientists rethink the "theory" so that the new information is taken into consideration. The theory must be retested and restated until all the known information can be supported by the theory. If the theory cannot be supported by the new information, then it must be thrown out.**

1.13 MEASUREMENTS

- In order to avoid confusion among scientists from different countries, standard measurements have been adopted for use in all scientific study. They are called the SI units.

Topic question

What are SI units? **SI stands for Systeme Internationale. SI units are the universal measurement units in scientific study.**

1.14 MICROSCOPY

- There are various types of microscopes, all of which magnify objects.

Topic question

What is the amount that a microscope can enlarge an object called? **Magnification**.

1.15 ADVANCED

- Electron microscopes are much more powerful than light microscopes and allow us to view tiny objects with excellent resolution.

Topic question

Which is more powerful, a light microscope or an electron microscope? **Electron microscopes are several hundred to a thousand times stronger than light microscopes.**

1.17 KEY CHAPTER POINTS

- Every organism which is alive shares the following properties: they are made up of one or more cells; they contain DNA; they reproduce; they are complex and organized; they are responsive to their environment; they extract energy from their surroundings; they maintain homeostasis; and they grow.

- All organisms are classified using the seven-level, six-kingdom, taxonomic system.

- Biology, as a science, is subject to the scientific method.

- SI units (i.e. the "metric system") are standardized units of measurements used in all scientific studies.

- Many instruments are used to make scientific observations. One of the most important in biology is the microscope.

2 The Composition and Chemistry of Life

2.0 CHAPTER PREVIEW

In this chapter we will:

- Understand how organisms obtain the energy they need to live and the relationship between photosynthesis and cellular respiration.

- Investigate the properties of the building blocks of all matter—atoms and molecules.

- Define isotopes and their properties.

- Investigate how covalent and ionic chemical bonds form by following the Law of Conservation of Mass.

- Discuss the unique properties of water that make it the best substance to surround and fill cells.

- Introduce basic concepts of solutes, solvents, acids, and bases.

2.1 BASIC ENERGY CONCEPTS

- All life forms require energy to live.

- Almost all organisms on earth obtain their energy from the carbohydrate molecule glucose.

- Plants make their own glucose by photosynthesis.

- Consumers obtain their glucose from the food they eat.

Topic question

How do plants make their energy molecules? **Photosynthesis.**

2.2 MATTER

- Matter is anything that takes up space. Matter can be seen and unseen (for example air is matter even though you cannot "see" air).

- Mass is different from matter. Mass is the quantity of matter an object has.

Topic question

Why are mass and matter not the same thing? **Matter is anything which takes up space. Mass is the amount of matter that an object or area contains. Students often confuse the words "mass" and "weight," but technically they do not mean the same thing. Weight is actually the amount of *force* exerted by an object. Weight is dependent on the force of gravity, but mass does not change. For example, if you took a bowling ball to the moon, it would weigh less on the moon than on earth. The mass of the bowling ball does not change. A bowling ball is a bowling ball. It contains the same amount of matter whether or not it is on the earth or the moon. But since there is more gravity on earth than the moon, the bowling ball exerts more force (i.e. weighs more) on earth than the moon.**

2.3 CHANGES IN MATTER

- Matter can be changed physically or chemically. An example of a physical change is when water goes from the physical form of ice to the physical form of water. A chemical change can only occur during a chemical reaction.

Topic question

Which type of change in matter occurs when water changes into ice? **Physical.** Which type of matter change occurs when gasoline is burned in a car? **Chemical.**

2.4 BASIC BUILDING BLOCKS OF MATTER

- Atoms are the basic units of matter. Pure forms of atoms are called elements.
- Atoms stick together to form bigger units called molecules.
- All organisms are made up of molecules.

Topic question

The great majority of the matter of organisms is in the form of atoms. True or False? **False. Almost all the matter of organisms consists of atoms linked to one another. Atoms linked to one another are called molecules.**

2.5 ATOMS AND ELEMENTS

- An element is matter composed of only one type of atom.
- The atom is made up of smaller units called subatomic particles. The three subatomic particles we will learn about are neutrons, protons, and electrons.

Topic question

What are the three subatomic particles which make up an atom? **Protons, neutrons, and electrons.** Which of these particles are in the nucleus? **Protons and neutrons.**

2.6 BOHR MODEL

- A Bohr model is a way to draw the nucleus and the electrons of an atom.

Topic question

When drawing a Bohr model, the electrons are placed inside of the nucleus. True or False? **False. When a Bohr model is drawn, the electrons are drawn outside of the nucleus, orbiting around it.**

2.7 PROTONS AND NEUTRONS

- There are two subatomic particles in the nucleus. They are the protons and neutrons.
- More than 95% of the mass of an atom is contained in the nucleus.

Topic question

Most of the space of a given atom is empty. True or False? **True. Although most of the mass of an atom is in the nucleus, most of the area taken up by an atom is empty space. This is because the electrons orbit so far outside of the nucleus that there is a large amount of "air" between the nucleus and the electrons.**

2.8 CHARGE OF SUBATOMIC PARTICLES

- Protons are positively charged.
- Electrons are negatively charged.
- Neutrons have no electrical charge.
- Normally, an atom has no electrical charge.

Topic question

Why does an atom normally not have an electrical charge? **Because normally the number of protons equals the number of electrons in an atom, which results in the atom having no overall electrical charge.**

2.9 SUBATOMIC PARTICLES ARE ALL THE SAME

- All protons are the same. All electrons are the same. All neutrons are the same. No matter what atom a given subatomic particle comes from, a given subatomic particle is the same. For example, an electron from a carbon atom is exactly the same as an electron from an argon atom as the same as an electron from a potassium atom.

Topic question

If all electrons are the same, why are there different atoms and elements? **Although all subatomic particles of a given type are all the same, it is the number of the subatomic particles one type of atom contains that makes it different from an atom of a different type.**

2.10 ATOMIC NUMBER AND ATOMIC SYMBOL

- The atomic symbol is a unique one or two letter designation given to all atoms.
- The atomic number is the number of protons an atom contains.

Topic question

What is the atomic symbol for calcium? **Ca. The student will need to look at the Periodic Table of the Elements to answer this.**

2.11 ATOMIC MASS

- The atomic mass of an atom-or element-can be nearly arrived at by adding the number of protons with the number of neutrons in the nucleus.

Topic question

What is the atomic mass of a hypothetical element with 27 electrons and 38 protons? **65. Add together 27 and 38.** What is the atomic mass of lead? **207. The student will need to look at the Periodic Table of the Elements for the answer.**

2.12 ELECTRON SHELLS

- Electrons orbit the nucleus in discreet and specific areas called shells, orbitals, or clouds.
- Electrons orbit the nucleus in pairs.
- When drawing a Bohr model, it is important to draw the electrons in their proper shell.

Topic question

How many electrons can the first shell hold? **Two.** How many electrons can the third energy shell hold? **Eighteen.**

2.13 ATOMIC PROPERTIES

- The properties of an atom are controlled mainly by the numbers of electrons an atom contains.
- Atoms that do not have filled electron shells are more reactive than ones which do.

Topic questions

What is a Noble gas? **Noble gasses have electron shells that are filled with paired electrons.**

What makes Noble gasses special? **Since the electron shells are filled, they are inert (they do not react well with other atoms).**

2.14 ISOTOPES

- An isotope is an element (atom) that has different forms because the different forms contain different numbers of neutrons.
- All isotopes of an element contain the same numbers of protons and electrons.

Topic question

You have identified what may be two new elements. One element contains 212 protons, 212 electrons, and 97 neutrons. The other element contains 212 protons, 212 electrons, and 208 neutrons. Have you indeed found two new elements? **No. Since both forms of this element contain the same number of protons and electrons, they are the same element. However, you have discovered two new isotopes of the same element since they contain different numbers of neutrons.**

2.15 MOLECULES

- Molecules are two or more atoms joined together.
- Molecules form when one atom shares its electrons with another atom. This is called a bond.
- Atoms tend to bond so that their electron levels are filled as a result of the bond.

Topic question

What is the octet rule? **The octet rule describes the fact that atoms are more stable when they have eight electrons in their outer electron shells. Most bonds form so that the octet rule is satisfied, which means that most atoms link together so that each has eight electrons in their outer shell following the bond.**

2.16 CHEMICAL BONDS ARE FORMED BY CHEMICAL REACTIONS

- A chemical bond is formed when two atoms share electrons.
- A chemical shorthand is used to describe chemical reactions.

Topic question

What happens when a chemical bond is formed? **Two atoms share their electrons, which causes the two atoms to stick together.**

2.17 COVALENT CHEMICAL BONDS

- Covalent bonds are formed when two atoms equally share the electrons participating in the bond.
- Covalent bonds are stronger than ionic bonds.
- Covalent bonds usually form so that the octet rule is satisfied. This results in both atoms involved in the bond obtaining a status of eight electrons in their outer most shell.

Topic question

In a covalent bond, the electrons are unequally shared by the atoms participating in the bond. True or False? **False.** There is no overall charge in a molecule with covalent bonds. True or False? **True. Since the electrons are equally shared there is no charge on the molecule.**

2.18 POLAR AND NON-POLAR COVALENT BONDS

- Polar covalent bonds are covalent bonds in which the electrons are not exactly evenly spread out between the atoms involved in the bond. This results in a slight "pull" of electrons toward one atom of the bond.
- Non-polar covalent bonds are formed when the electrons are almost equally shared between atoms involved in the bond.

Topic question

Describe why a molecule that forms by polar covalent bonds, such as water, forms such a bond. **There is a slight negative charge on one end of the molecule and a slight positive charge on the other end of the molecule because the oxygen atom has many more protons in the nucleus than the hydrogen atoms do. Since electrons are negatively charged and protons are positively charged, this results in the electrons of the covalent bond being slightly more attracted to the oxygen nucleus than the hydrogen nuclei. Since the negatively charged electrons are closer to one end of the molecule than the other, a slight negative charge develops there, and a slight positive charge develops opposite the area where the electrons are spending the most time.**

2.19 IONIC BONDS

- An ion is an atom that has a negative or a positive charge.

- An ionic bond forms between a positively charged ion and a negatively charged ion.

- A negatively charged ion is an electron acceptor, and a positively charged ion is an electron donor.

Topic question

Why is an electron acceptor negatively charged? **An electron acceptor receives a donated electron. Since electrons are negatively charged, accepting an electron means that an atom gains one more electron than it is supposed to have. This results in one more negatively charged particle for an electron, which equates to a negative charge.**

2.20 REACTANTS AND PRODUCTS

- Molecules or atoms written on the left side of a chemical reaction are called reactants.

- Molecules or atoms written on the right side of a chemical reaction are called products.

Topic question

What are reactants and products? **Reactants are the atoms or molecules that "go into" (or participate in) a chemical reaction, and products are what "comes out of" (or results from) a chemical reaction.**

2.21 LAW OF CONSERVATION OF MASS

- The Law of Conservation of Mass is always followed when a chemical reaction takes place.

- The Law states that the amount of matter which enters into a reaction must be equal to the amount of matter which comes out of a reaction.

- The earth is considered a closed system, which is why the Law of Conservation of Mass is true.

Topic Question

In a closed system, there is a certain acceptable "loss" of matter during chemical reactions. True or False? **False. According to the Law of Conservation of Mass, mass can neither be destroyed ("lost") nor created during a chemical reaction. What goes in must equal what comes out.**

2.22 WATER

- Water has many unique properties, which has resulted in it being critical to the functioning of all life forms.

- The uniqueness of water is because of the way in which the molecule is formed.

- The bonding interaction between oxygen and hydrogen in water results in a polar covalent bond.

- Water forms hydrogen bonds between molecules because of the polar nature of the covalent bond.

- The molecular structure of water allows molecules to fit into one another tightly.

- Water holds it temperature well.

Topic question

What are the properties of water which make it unique to support life? **Water is polar, forms hydrogen bonds, is a non-linear molecule, and is slow to cool or warm.**

2.23 SOLUTIONS, SOLVENTS, AND SOLUTES

- A solution is a homogeneous mixture of a gas, liquid, or a solid.

- A solvent is that part of a solution which is in higher quantities.

- A solute is that part of a liquid which is in lower quantities.

Why is it important to understand the concepts of solutions, solvents, and solutes? **All metabolic processes of life occur in solution environments. In all these environments, the solvent is water.**

2.24 ACIDS AND BASES

- An acid is a solution that has more hydrogen ions than hydroxide ions.
- A base is a solution that has more hydroxide ions than hydrogen ions.

Topic question

What makes one acid stronger than another? **The more hydrogen ions present in an acid solution, the stronger the acid will be.**

2.25 pH SCALE

- The pH scale is used to measure the acidity or basicity of a solution.

Topic question

Does an alkaline solution have a low or high pH? **Basic, or alkaline, solutions have more hydroxide ions in them than hydrogen ions. They have a higher pH than acidic solutions.**

2.26 INDICATORS

- Indicators are chemicals that can be added to a solution and change different colors based on the pH of the solution.

Topic question

List two ways that the pH of a solution can be measured. **With an indicator or a pH meter.**

2.28 KEY CHAPTER POINTS

- All life forms require energy to live.
- Everything that exists—living and nonliving—is composed of matter. This matter can undergo physical and chemical changes.
- Atoms are the smallest units of matter. They are composed of protons, neutrons, and electrons.
- Protons and neutrons make up the nucleus of the atom. Electrons orbit the nucleus in shells (also called clouds and orbitals).
- Atoms can link to one another by covalent or ionic bonds to form larger molecules. This occurs through chemical reactions, which follow the Law of Conservation of Mass.
- Water is special because it is a polar molecule that is non-linear, and hydrogen bonds with other water molecules. This gives water unique properties, making it critical in the maintenance of life.
- The pH scale measures acidity of solutions.

Basic Biochemistry of the
3 Molecules of Life

3.0 CHAPTER PREVIEW

In this chapter we will:

- Define what organic molecules are and why they are important to life.

- Understand the difference between organic and inorganic molecules.

- Learn about isomers and why they require the use of structural formulae to more accurately depict organic molecules.

- Investigate how the four classes of organic molecules—carbohydrates, proteins, lipids, and nucleic acids—are made and broken down by chemical reactions in the cell.

3.1 OVERVIEW

- Most of the matter of all organisms is composed of water. The second largest composition of matter are organic molecules.

Topic questions

Organic molecules are synthesized by all living organisms. True or False? **True.**

3.2 ORGANIC MOLECULES

- Organic molecules are also called carbon compounds because all organic molecules contain carbon.

- The four classes of organic molecules are fats, carbohydrates, proteins, and nucleic acids.

Topic question

What four types of molecules do organic chemists study? **Proteins, fats (lipids), carbohydrates, and nucleic acids.**

3.3 ORGANIC VS. INORGANIC MOLECULES

- Organic molecules are large, always bond covalently, and contain carbon. Inorganic molecules are usually small (have a few atoms), bond ionically, and do not contain carbon.

- Molecules with bonds between carbon and hydrogen are organic molecules.

Topic question

List the differences between organic and inorganic molecules. **Organic molecules have bonds between carbon and hydrogen. Also, organic molecules are usually much larger than inorganic molecules and bond covalently.**

3.4 MOLECULAR FORMULAE AND ISOMERS

- A molecular formula is not specific enough to describe organic molecule bonding structures. Therefore, the structural formula is used.

- Because there are many different types of bonds that carbon can form, organic molecules easily form isomers. An isomer is made up of two or more molecules with the same molecular formulae but different structural formulae.

- It is important to realize that even though isomers have the same molecular formulae, their physical properties are different because their structures are different.

Topic questions

What is an isomer? **Two or more molecules that share the same molecular formula, but have different structures.**

Why are isomers important? **Since isomers share the same molecular formula, it is important to realize that the reason isomers behave differently is because their structures are different. Structural formulae account for this difference in structures that isomers have, which is the reason that organic chemistry uses structural rather than molecular formulae.**

3.5 MOLECULES ARE 3-D

- All molecules are three-dimensional.

Topic question

Are molecules as two-dimensional as they appear when their structural formula is written on a piece of paper? **No. All molecules are three-dimensional. This is often a difficult concept for students to grasp. Sometimes it helps for them to take two pencils and lay one on the table to make a two-dimensional structure. It has width and length on the table. Then take the other pencil and place it with the eraser on top of the pencil on the table with the point extending upwards. Now, the two pencils, which represent one molecule, form a three-dimensional structure.**

3.6 PHYSICAL PROPERTIES OF ISOMERS ARE DIFFERENT

- The three-dimensional structure of isomers is what makes them have different physical and chemical properties.

Topic question

Why do isomers behave differently? **Isomers have different structures. A molecule's physical and chemical properties are dependent on the structure of the molecule.**

3.7 BONDING PATTERNS

- Carbon can form single, double, or triple bonds with other atoms.

Topic question

How many different types of bonds can carbon form? **Three.**

3.8 CARBON IS THE SPECIAL ATOM

- Because of a wide range of properties, carbon is uniquely suited to be the structural backbone atom of all organic molecules.

Topic question

Why is carbon uniquely suited to be the organic molecule backbone atom? **Carbon can bond with four other atoms. Also, it can accommodate a variety of shapes—linear, branched, and ring—allowing for many different molecule shapes and sizes.**

3.9 ORGANIC REACTIONS

- Organic reactions are used by life forms to synthesize organic molecules.
- There are two basic types of organic reactions: anabolic and catabolic.

Topic question

What is the difference between anabolic and catabolic reactions? **Anabolic reactions synthesize organic molecules and catabolic reactions break organic molecules down into smaller units.**

3.10 POLYMERS ARE BUILT FROM MONOMERS

- Organic molecules are built from small units called monomers into larger units called polymers.

Topic questions

What is a monomer? **A monomer is the basic building block of all organic molecules.**

3.11 DEHYDRATION SYNTHESIS REACTIONS

- A dehydration synthesis reaction is one in which water is formed when two monomers are linked together. Since water is a product of the reaction, the term "dehydration" is used to describe these anabolic reactions.

- Repeated dehydration synthesis reactions link together hundreds to thousands of monomers together to make a polymer.

Topic question

How are polymers made? **Polymers are made by linking together hundreds to thousands of monomers through dehydration synthesis reactions.**

3.12 HYDROLYSIS REACTIONS

- Hydrolysis reactions are catabolic reactions that break polymers into smaller monomer units by removing one monomer at a time from the polymer.

Topic question

Why is a hydrolysis reaction so named? **Because water is used to break the bond between the monomer and the polymer. Since water is added to the bond, the name hydrolysis or hydration reaction is used.**

3.13 CARBOHYDRATES: THE "SACCHARIDES"

- A saccharide is a sugar (carbohydrate) molecule.

Topic question

How can you tell if a molecule is a carbohydrate? **Usually most carbohydrate molecule names end in the letters "-ose." (glucose, lactose, etc.)**

3.14 MONOSACCHARIDES ARE THE MONOMERS FROM WHICH CARBOHYDRATE POLYMERS ARE BUILT

- Polymers of carbohydrates are built from monosaccharides.

Topic question

From what molecules are carbohydrates built? **Carbohydrates are polymers of monosaccharide units.**

3.15 MONOSACCHARIDE ISOMERS AND CONFORMATION

- Monosaccharides can take on two structural shapes: linear (or straight chain) and ring (or cyclic).

Topic question

What conformation (conformation means structure) do monosaccharides usually take? **Usually monosaccharides are in the ring (cyclic) form.**

3.16 POLYSACCHARIDE SYNTHESIS AND GLYCOSIDIC BONDS

- A glycosidic bond is formed by a dehydration synthesis reaction. It links two saccharide units together.

Topic question

What is a glycosidic bond? **A glycosidic bond is a bond that links together two saccharides. It forms between a carbon of one saccharide, an oxygen molecule, and a carbon atom of the other saccharide.**

3.17 DISACCHARIDES

- A disaccharide is a molecule formed from two monosaccharides. It is formed by a dehydration-synthesis reaction, which forms a glycosidic bond between the two monosaccharides.

Topic question

How is a disaccharide held together? **Through a glycosidic bond between the two monosaccharide units of the disaccharide.**

3.18 POLYSACCHARIDES

- Polysaccharides are formed when three or more monosaccharides are linked together.

Topic question

Describe the process by which a polysaccharide is synthesized. **Monosaccharides are linked together repeatedly by glycosidic bonds to form a polysaccharide.**

3.19 CARBOHYDRATE MACROMOLECULES

- Starch, glycogen, and cellulose are common forms of carbohydrate macromolecules.

Topic question

What is the most abundant organic molecule on the planet? **Cellulose.**

3.20 LIPIDS (FATS)

- Lipids are formed from carbon, hydrogen, and oxygen atoms, but in a much different ratio than carbohydrates.

Topic question

What type of molecules are steroids, oils, and waxes? **They are organic molecules; specifically lipids (fats).**

3.21 FATTY ACIDS

- A fatty acid is a long carbon chain with a carboxylic acid group attached to one end.
- There are two types of fatty acids: saturated and unsaturated.

Topic question

What is the difference between a saturated and an unsaturated fatty acid? **Saturated fatty acids contain many more hydrogen atoms attached to the carbons—the carbon chain is saturated with hydrogen. Also, saturated fatty acids do not have double bonds between the carbons, whereas unsaturated fatty acids do.**

3.22 GLYCEROL BACKBONE AND ESTER BONDS

- Glycerol forms the backbone of all lipids.
- Fatty acids are linked to glycerol by ester bonds.

Topic question

What is an ester bond? **A bond that holds a fatty acid to the glycerol backbone. Specifically, an ester bond is an oxygen atom bonded to a barb on the glycerol and a carbon on the fatty acid, linking the two together.**

3.23 PROTEINS

- Proteins are structurally diverse molecules. Not only do they contain carbon, hydrogen, and oxygen, but also other atoms such as sulfur and nitrogen.
- Proteins are usually large molecules.

Topic question

What are two functions which proteins serve? **They are important structural molecules and enzymes.**

3.24 AMINO ACIDS ARE THE MONOMERIC UNITS OF PROTEINS

- The monomeric unit of proteins are called amino acids.

Topic question

What is the monomeric unit of a protein? **An amino acid.**

3.25 PEPTIDE BONDS

- A peptide links two amino acids together. A peptide bond forms between a nitrogen from one amino acid and a carbon from the other amino acid.

Topic question

What is an amino bond? **A bond which holds two amino acids together.**

3.26 DI– AND POLYPEPTIDES

- Larger and larger proteins are formed through repeated dehydration synthesis reactions in which amino acids are linked together.

Topic questions

What is a dipeptide? **Two amino acids linked together.**

What is a polypeptide? **Many amino acids linked together.**

3.27 NUCLEIC ACIDS

- Nucleic acids are organic molecules that store all the information an organism needs to know to survive.

Topic question

3.28 NUCLEOTIDES

- Nucleotides are the monomeric units of nucleic acids.
- Repeatedly linking together nucleotides by dehydration synthesis reactions forms a nucleic acid.

Topic question

What are the three components of a nucleotide? **A central five-carbon sugar (called a pentose), a phosphate group, and a nitrogen containing base (called a nitrogenous base).**

3.29 DNA, OR DEOXYRIBONUCLEIC ACID

- DNA is the nucleic acid that contains the genetic information. The nucleotides of DNA all contain a phosphate group and the sugar deoxyribose. There are one of four nitrogen bases that a nucleotide of DNA may have: thymine, guanine, adenine, and cytosine.

Topic question

What are the three components of the nucleotides of DNA? **Deoxyribose is the pentose. The nitrogenous bases are either thymine, adenine, guanine, or cytosine. All the nucleotides of DNA contain a phosphate group as well. (Remind the student that all nucleotides contain a phosphate group.)**

3.30 RNA, OR RIBONUCLEIC ACID

- RNA is a messenger nucleic acid. It is also made from four nucleotides which are linked together repeatedly.

Topic question

What are the four nitrogen containing bases that make up the nucleotides of RNA? **Uracil, cytosine, adenine, and guanine.**

3.31 ATP, OR ADENOSINE TRIPHOSPHATE

- ATP is a special nucleotide composed of a central ribose molecule with the nitrogen base adenine attached to one end and three phosphate groups attached to the other end.

Topic question

What makes ATP a special molecule? **The three phosphate groups.**

3.33 KEY CHAPTER POINTS

- Organic molecules always contain carbon and hydrogen, and often contain oxygen.
- Because of isomers, writing the structural formula for a molecule is preferred over the simple chemical (molecular) formula.
- Carbon has many unique properties, making it specially suited to serve as the main atom in organic molecules.
- Polymers are built from monomers using dehydration (or condensation) synthesis reactions.
- Polymers can be broken into their smaller monomeric units using hydrolysis reactions.
- Carbohydrates are built from monosaccharide monomers through the glycosidic bond.
- Lipids are built by linking as many as three fatty acids to a backbone molecule of glycerol using the ester bond.
- Proteins are built from amino acid monomers through the peptide bond.
- Nucleic acids are built from nucleotides.

Introduction to the
4 | Cell and Cell Membrane

4.0 CHAPTER PREVIEW

In this chapter we will:

- Review "cell theory."
- Discuss the general features and functions of a cell.
- Define the two basic cell types—prokaryotic and eukaryotic.
- Review basic cell structure.
- Investigate the structure and function of the outer boundary of the cell—the cell membrane.
- Understand why cells are "so small."

Topic question

4.1 OVERVIEW

- This chapter will present the fact that cells are the basic unit of life as well as explore the function of the cell membrane.

Topic question

4.2 CELL THEORY

- Cell theory is the theory that the cell is the basic functional unit of life. All life forms are composed of one or more cells.

Topic question

What is the study of the cell called? **Cytology.**

4.3 GENERAL FEATURES OF CELLS

- Cells are diverse in shape, size, and function.

Topic question

What is a group of cells called that all perform the same function? **A tissue.** What is a group of tissues that all have the same function called? **An organ.**

4.4 UNICELLULAR AND MULTICELLULAR ORGANISMS

- An organism composed of only one cell is a unicellular organism.
- An organism composed of more than one cell is a multicellular organism.

Topic question

What is a unicellular organism? **An organism composed of only one cell.**

4.5 COMMON STRUCTURES OF ALL CELLS

- All cells have a cell membrane. Also, all cells contain DNA on their interior along with organelles.

Topic question

What are the units inside of cells that perform certain specific functions called? **Organelles.**

4.6 EUKARYOTIC CELLS

- Eukaryotic cells contain membrane-bound organelles. They also have their DNA housed in a special compartment within the cell interior called the nucleus.

Topic question

What features distinguish a eukaryotic cell? **A nucleus and membrane-bound organelles.**

4.7 PROKARYOTIC CELLS

- Prokaryotic cells do not have a nucleus, nor do they have membrane-bound organelles.

Topic question

If you were standing inside of a cell and saw the DNA free-floating in the cytoplasm, in what type of cell would you be standing? **Prokaryotic. Since the DNA is free in the cytoplasm, you know you are not in a eukaryotic cell because the DNA of eukaryotic cells is contained in the nucleus.**

4.8 FUNCTIONS OF A CELL: REPRODUCTION

- All cells reproduce.

Topic question

Since bacteria are unicellular, they do not reproduce. True or False? **False. Bacteria are living organisms and all living organisms reproduce whether they are unicellular or multicellular.**

4.9 FUNCTIONS OF A CELL: ABSORPTION

- Cells bring into, or absorb, essential substances.

Topic question

The cell structure that has the most to do with determining what gets into the cell is _____. **The cell membrane.**

4.10 FUNCTIONS OF A CELL: DIGESTION/PROCESSING

- Digestion is the process of breaking down substances brought into the cell.

Topic question

What happens to a large molecule when it is digested? **It is broken down into smaller molecules.**

4.11 FUNCTIONS OF A CELL: BIOSYNTHESIS

- All cells make substances out of absorbed molecules. Biosynthesis is the process of synthesizing molecules from substances which are absorbed.

Topic question

What is biosynthesis? **The process of making molecules out of absorbed molecules.**

4.12 FUNCTIONS OF A CELL: MOVING SUBSTANCES OUT OF THE CELL

- There are three processes that the cell uses to move things out of the cell—excretion, egestion, and secretion.

Topic question

What is excretion? **The process of moving waste products that are soluble in water out of the cell.**

4.13 FUNCTIONS OF A CELL: RESPOND TO THE ENVIRONMENT

- All cells respond to changes in their environments.

Topic question

What is irritability? **The property of a cell to respond to its environment.**

4.14 FUNCTIONS OF A CELL: MAINTAIN HOMEOSTASIS

- All cells maintain a stable internal environment, a process called homeostasis.

Topic question

Why is it important for a cell to maintain homeostasis? **If a stable environment is not maintained, then the cell cannot function optimally.**

4.15 FUNCTIONS OF A CELL: PROTECTION

- The cell membrane (and the cell walls in organisms that have one) protects the cell from the environment.

Topic question

Animal cells are protected from the environment by their cell membrane and cell wall. True or False? **False. Animal cells do not have cell walls.**

4.16 INSIDE THE CELL

- The interior of the cell contains a jelly-like fluid called cytoplasm. Within the cytoplasm are the various organelles.

Topic question

Where is the DNA of prokaryotes found? **Free in the cytoplasm.**

4.17 REVIEW OF CELL STRUCTURE

- All cells have a cell membrane, cytoplasm, organelles, and DNA.
- Some cells have a layer outside the cell membrane called a cell wall.

Topic question

What distinguishes prokaryotes from eukaryotes from a DNA standpoint? **Prokaryotic DNA is not contained in a nucleus, but eukaryotic DNA is.**

4.18 CELL MEMBRANE

- The cell membrane is made up mostly of molecules of lipid and protein.
- The cell membrane regulates which molecules can get in and out of the cell.

Topic question

What is the cell membrane? **The outer boundary of the cell. It regulates what substances get into and out of the cell.**

4.19 PHOSPHOLIPID

- The cell membrane is composed of a special type of lipid called phospholipid.
- A phospholipid is unique because of its structure. It has one end which is hydrophilic (the phosphate-group end) and one end which is hydrophobic (the fatty-acid chain end).

Topic question

See next section

4.20 LIPID BILAYER

- The cell membrane is oriented as a two-layer structure called the lipid bilayer.

Topic question

What is meant by the lipid bilayer? **The membrane of the cell forms into two layers of phospholipids because of the structure of the phospholipid molecule. The hydrophobic tails cluster together in the middle of the bilayer and the hydrophilic heads protrude into the water matrix.**

4.21 PROPERTIES OF LIPID BILAYER

- The lipid bilayer is semi-permeable.

Topic question

What is meant by semi-permeable? **Semi-permeable refers to the fact that the cell membrane allows only certain substances to pass freely through. This effectively limits what goes in and out of the cell, protecting it.**

4.22 MEMBRANE PROTEINS

- There are two types of membrane proteins embedded in the lipid bilayer—peripheral and integral proteins.

- Membrane proteins have two important functions—to serve as anchor points for the attachment of other proteins inside and outside of the cell and to serve as channels for molecules to pass through the membrane.

Topic question

What is the relationship between semi-permeability and membrane proteins? **The membrane proteins that serve as channels for substances to pass through are one of the reasons the membrane is semi-permeable. Very few molecules can just pass through the membrane without passing through a channel protein. If the channel protein does not allow the substance to pass through, it does not get into or out of the cell.**

4.23 FLUID MOSAIC MODEL

- The fluid mosaic model of the cell membrane explains the way in which the membrane and membrane proteins have been observed to act over the past fifty years. It states that the cell membrane is not rigid but fluid. The membrane can move like the walls of a water balloon do. The mosaic part refers to the movement of proteins and other molecules within the cell membrane from one place to another in the membrane.

Topic question

Why can proteins move around within the cell membrane? **The cell membrane is not rigid. It allows the molecules within it to move freely through the membrane. Also, the fact that the protein molecules have hydrophobic portions allows that to be possible. This is referred to as the mosaic nature of the cell membrane.**

4.24 CELLULAR TRANSPORT

- Cellular transport is the general name of the process by which molecules and substances move across the cell membrane into or out of the cell.

Topic question

What is cellular transport? **The process of moving molecules/substances/"things" into or out of the cell across the cell membrane.**

4.25 DIFFUSION

- Diffusion is the passive movement of a molecule down its concentration gradient through random molecule movement. The overall movement of molecules in diffusion is from an area of higher to an area of lower concentration.

- Some molecules can pass into and out of the cell by diffusion.

Topic question

What are two molecules that can pass through the cell membrane by diffusion? **Oxygen and carbon dioxide.**

4.26 GATED CHANNELS

- Gated channels are integral membrane proteins that allow molecules to pass through the membrane by diffusion.

Topic question

Why is gated channel transport a type of passive transport? **Because the cell does not need to expend any energy to perform gated channel transport, it is a type of passive transport. Also, the movement of the molecules through their respective gated channels occurs randomly down the concentration gradient, which is the definition of diffusion.**

4.27 OSMOSIS

- Osmosis is a special type of diffusion that relates to the movement of a solvent across a semi-permeable membrane. In osmosis, the membrane is permeable only to the solvent. The solvent moves randomly across the membrane from the area of high solvent concentration to the area of lower solvent concentration.

- In biological systems, the most common solvent that is moved through osmosis is water.

Topic question

Describe what happens to a cell when it is in a hypertonic environment. **This is a biological system, so the solvent is water. Since the cell is in a hypertonic environment, the concentration of water inside the cell is higher than the concentration of water outside of the cell. The solvent will move across the membrane from the area of higher solvent concentration to the area of lower solvent concentration. This means that water will move from the inside of the cell to the outside of the cell. This results in the cell getting all shriveled up since it loses a lot of water in this situation. Yes, this really is a concern in medicine. People have been killed due to too rapid infusion of hypertonic fluid in their veins when they are extremely dehydrated.**

4.28 REVIEW OF PASSIVE TRANSPORT MECHANISMS

- The modes of passive transport are diffusion, gated channel transport, and osmosis.

Topic question

4.29 ACTIVE TRANSPORT

- Active transport requires the cell to use energy to move molecules. One of the reasons why energy must be expended is because active transport often moves molecules up their concentration gradient.

Topic question

How is active transport accomplished through membrane proteins? **The membrane protein serves as a channel through which the molecule can be moved up its concentration gradient. The cell pumps the molecule through the protein channel, using energy to do so.**

4.30 CYTOSIS

- Endocytosis moves substances into the cell via a vesicle. Exocytosis moves substances out of a cell via a vesicle.

Topic question

Describe the process of exocytosis. **The cell forms a membraned vesicle around the substance to be moved out of the cell. Then the vesicle is moved through the cytoplasm to the cell membrane. The membrane of the vesicle merges with the cell membrane. As it does, the inside of the vesicle is exposed to the outside of the cell, and the material** inside of the vesicle is expelled outside of the cell. This process requires energy and is a mode of active transport.

4.31 ADVANCED

- Cells are so small because of the surface to volume ratio.

Topic question

Why is there a limit to how large a cell can be? **An answer of "because of the surface to volume ratio" is not enough. When the cell increases in size, both the surface area and the volume of the cell increase in size. However, the volume of the cell increases more rapidly than the surface area of the cell. The stuff inside the cell is dependent on the surface area of the cell to feed it. That is, things get into and out of the cell across the cell membrane. So there comes a point at which the surface area of the cell cannot keep up with the demands of the inside of the cell. Nutrients cannot get into the cell and wastes cannot get out of the cell fast enough, and the cell dies. This is why there is a limit to how large a single cell can be.**

4.33 KEY CHAPTER POINTS

- Cell theory states that cells are the basic functional unit of all life.

- Organisms that contain only one cell are unicellular organisms. Cells that contain more than one cell are multicellular organisms.

- All cells:
    - have DNA
    - have diverse shapes and functions
    - have a cell membrane to protect it and organelles inside
    - reproduce themselves
    - can absorb nutrients from their environments
    - can digest and process nutrients
    - can synthesize organic molecules
    - can move substances into, out of, and throughout the cell
    - can respond to the environment
    - can maintain homeostasis

- There are two basic cell types—prokaryote and eukaryote.

- The cell membrane protects the cell from the environment.

- The cell interior contains the organelles and DNA.

- The cell membrane is a two-layer structure composed of phospholipids and proteins.

5 | The Cell Interior and Function

5.0 CHAPTER PREVIEW

- Investigate and understand the organization and function of the cell interior.

- Define the differences between eukaryotic and prokaryotic cell structure.

- Discuss the structure and function of the following eukaryotic organelles and structures:
  - Protoplasm
  - Cytoplasm
  - Nucleoplasm
  - Cytoskeleton
  - Nucleus
  - Ribosome
  - Endoplasmic reticulum
  - Golgi apparatus
  - Lysosomes and peroxisomes
  - Mitochondria
  - Plastids
  - Vacuoles
  - Middle lamella
  - Extracellular matrix

5.1 OVERVIEW

- This chapter focuses on the interior cell structure and function.

Topic question

Which cell type is structurally more complex, eukaryotic or prokaryotic? **Eukaryotic. They have many more organelles than prokaryotic cells.**

5.2 ORGANELLES—GENERAL

- Organelles are units contained inside a cell that carry out specific functions.

Topic question

What are the organelles and intracellular substances discussed in this chapter? **Protoplasm, cytoplasm, nucleoplasm, nucleus, nucleolus, cytoskeleton, ribosome, endoplasmic reticulum, Golgi apparatus, mitochondria, lysosome, peroxisome, vacuoles, plastids, centrioles.**

5.3 PROTOPLASM

- The protoplasm refers to all the substances inside of a cell. Protoplasm is basically everything inside the cell membrane, including the organelles, cytoplasm, and nucleus.

- Cytoplasm is everything inside the cell, but outside of the nucleus. The cytoplasm does not include the nucleus.

- Cytosol is the fluid component of the cytoplasm. It has the consistency of jelly.

Topic questions

Does the cytoplasm include the nucleus? **No. The cytoplasm includes everything inside the cell except the nucleus.** What is the cytosol? **It is the aqueous part of the cytoplasm. It is about 70% water and 30% ions and proteins.**

5.4 NUCLEOPLASM

- The nucleoplasm is the name of the interior substance of the nucleus.

Topic question

What is contained in the nucleoplasm? **Ions, water, proteins, and DNA.**

5.5 CYTOSKELETON

- The cytoskeleton is a complex meshwork of proteins which maintains the structure of the cell.

Topic question

What functions does the cytoskeleton serve? **It maintains the shape (structure) of the cell. It also serves as an anchoring point for organelles and as the pathway for organelles and materials to be transported in the cell.**

5.6 CYTOSKELETAL PROTEINS, CILIA, AND FLAGELLA

- The cytoskeletal proteins are microtubules, intermediate fibers, and microfilaments.
- Cilia and flagella are composed of microtubules.

Topic question

What is the main function of cilia and flagella? **Cilia and flagella are composed of microtubules. They are both mobile; that is they move. Cilia and flagella both beat to move organisms in their environment or to move material past the organism or cell. Therefore, the main function of cilia and flagella is involved in transport.**

5.7 EUKARYOTIC ORGANELLES—GENERAL

- Eukaryotic cells contain more organelles than prokaryotic cells.

Topic question

5.8 NUCLEUS

- The nucleus contains DNA and nucleoplasm.
- The nuclear membrane is a double-membrane structure surrounding the nucleus.
- Nuclear pores are holes within the nuclear membrane that allow substances to pass through.
- The nucleoplasm, nuclear lamina, DNA (in the form of chromosomes), and nucleolus are contained in the nucleus.

Topic questions

Where is the nucleus found? **Inside all eukaryotic cells. Prokaryotic cells do not have a nucleus. Further, the nucleus is found within the cytoplasm of the cell, although it is not considered part of the cytoplasm.**

How many lipid bilayers make up the nuclear membrane? **There are two lipid bilayers which make up the nuclear membrane. This means there are four individual lipid layers in the nuclear membrane.**

What is the function of the nucleolus? **The nucleolus manufactures ribosomal pieces that are transported out of the nucleus and assembled into ribosomes in the cytoplasm.**

What is the nuclear lamina? **The nuclear lamina is the skeleton of the nucleus. It is composed of proteins called intermediate fibers and maintains the structure of the nucleus.**

Christian perspective

Be in this world not of this world.

5.9 RIBOSOMES

- Ribosomes are found in the cytoplasm and manufacture proteins.

Topic question

What is a polysome? **A polysome is a collection of many ribosomes in the cytoplasm.**

5.10 ENDOPLASMIC RETICULUM (ER)

- Endoplasmic reticulum is made up of folded layers of membranes in the cytoplasm. ER is important in protein modification and transport.

- ER with ribosomes attached to it is called rough ER. ER without attached ribosomes is called smooth ER.

Topic question

What is the structure and function of ER? **ER is a series of folded membranes connected directly to the nuclear membrane. ER functions to modify and transport proteins/ molecules within the cell.**

5.11 GOLGI APPARATUS AND VESICLES

- The Golgi apparatus is also a series of folded membranes. The Golgi further modifies, stores, and packages proteins after they are made. Golgi also makes lysosomes and peroxisomes.

- A vesicle is formed by the ER and the Golgi to transport certain molecules in the cell. A vesicle is a membrane-bound compartment into which molecules and other substances are placed for transport or digestion.

Topic question

What is a signal group and why is it important? **A signal group is a short series of molecules added to a protein that tells the cell where the protein should be taken after it is made. If there were no signal groups added by the Golgi, then the cell would not know where to take proteins after they have been made.**

5.12 LYSOSOMES AND PEROXISOMES

- Lysosomes and peroxisomes are organelles that contain digestive enzymes to break down various substances the cell absorbs or needs to process.

Topic question

Where are lysosomes made? **The Golgi apparatus.** Where are peroxisomes made? **They are self-assembling in the cytoplasm.**

5.13 MITOCHONDRIA

- Mitochondria are organelles that manufacture the cell's energy molecule, ATP. Like the nucleus, this is also a double membrane-bound organelle.

Topic question

What are cristae? **Cristae are the folds of the inner membrane of mitochondria.**

5.14 PLASTIDS

- Plastids are organelles found only in plants. The plastid called a chloroplast performs photosynthesis. Other plastids function to store substances.

Topic question

How many membranes do plastids have? **Like mitochondria and the nucleus, plastids are double membrane-bound organelles. This means they have two lipid bilayers, or four individual layers of membrane.**

5.15 VACUOLES

- Vacuoles are storage organelles that are membrane bound.

Topic question

What is the function of a vacuole? **Vacuoles function to store substances within the cell.**

5.16 HOLDING PLANT CELLS TO ONE ANOTHER—THE MIDDLE LAMELLA

- The middle lamella is a sticky layer of polysaccharides secreted by plant cells that holds plant cells to one another.

Topic question

Where is the middle lamella found and what is its function? **The middle lamella is found only in plants. It is a layer secreted outside the plant cell walls and holds plant cells together.**

5.17 HOLDING ANIMAL CELLS TO ONE ANOTHER—THE EXTRACELLULAR MATRIX (ECM)

- The extracellular matrix is found only in animals. It serves to hold animal cells to one another. It is composed of the glycoprotein collagen, proteoglycans, and fibronectin.

Topic question

If you were standing on the outside of a cell and found that the molecules around you were fibronectins, collagen, and proteoglycans, would you be standing in a plant or animal organism? **Animal. Plant cells are held together by sticky sugar molecules called polysaccharides.**

5.18 PROKARYOTE CELL STRUCTURE

- Prokaryotic cell structure will be covered more in the bacteria chapters. What is important to understand now is that prokaryotic DNA is not contained in a nucleus. Prokaryotes have no nucleus. Their DNA is free in the cytoplasm

Topic question

What is the nucleoid? **The area of the cytoplasm where prokaryotic DNA is stored.**

5.20 KEY CHAPTER POINTS

- Organelles are special structures inside of cells that perform specialized functions.
- The inside of a cell is filled with a jelly-like substance called cytoplasm, and the inside of the nucleus contains nucleoplasm.
- The cytoskeleton holds the cell together.
- The nucleus is surrounded by a double lipid bilayer with holes in it, called pores. The nucleus contains the DNA and nucleolus.
- Mitochondria, ribosomes, endoplasmic reticulum, Golgi apparatus, vesicles, vacuoles, peroxisomes, and lysosomes are organelles found in all eukaryotic cells. All of them are bound by a membrane except ribosomes.
- Plants have plastids, a type of organelle not found in animal cells.
- Animal cells have centrioles, a type of organelle not found in plant cells.
- Almost all plant cells are also surrounded by a cell wall. No animal cells have cell walls.
- Animal cells are held to one another by the extracellular matrix.
- Plant cells are held to one another by the middle lamella.

Metabolism Overview
6 and Enzymes

6.0 CHAPTER PREVIEW

In this chapter we will:

- Define types of metabolism.

- Investigate the properties of potential and kinetic energy.

- Understand that living things convert one form of energy to the other to stay alive.

- Review the First and Second Law of Thermodynamics.

- Investigate the activation energy-lowering and energy-coupling properties of enzymes.

- Classify types of enzymes based on the chemical reactions they perform.

- Define cofactors and coenzymes.

6.1 OVERVIEW

- All organisms require energy to stay alive.

Topic question

What is ATP? **The molecule that almost all organisms use for energy (this will be covered much more in Chapter 8).**

6.2 METABOLISM

- The summation of all the chemical reactions in an organism is called metabolism.

Topic question

What is metabolism? **The summation or totality of all chemical processes (or chemical reactions) occurring in an organism.**

6.3 CATABOLIC AND ANABOLIC METABOLISM

- Anabolic processes are chemical reactions that synthesize organic molecules or store energy.

- Catabolism are chemical reactions that break molecules down or release energy.

Topic questions

What is a typical anabolic reaction that we have learned about? **Condensation synthesis (or dehydration synthesis) reactions.**

What is a typical catabolic reaction we have learned about? **Hydrolysis (or hydration) reactions.**

6.4 POTENTIAL AND KINETIC ENERGY

- Potential energy is the ability to do work.

- Kinetic energy is active work being done or energy of motion.

Topic question

What is the difference between chemical and mechanical energy? **Chemical potential energy is energy stored within chemical bonds of molecules. Chemical kinetic energy is work done as the result of release of energy from a chemical reaction. Mechanical potential energy is energy stored in an object. Mechanical kinetic energy is energy transferred from one object to another.**

6.5 ENERGY CONVERSION

- The conversion of energy from potential to kinetic chemical energy is what allows life forms to survive.

Topic question

When heat is given off by the catabolism of glucose, has energy been converted? If so, how? **Yes, energy has been converted. The potential energy stored in the bonds of the glucose molecule has been released in the form of heat. Heat is a form of kinetic energy.**

6.6 BASIC THERMODYNAMICS

- Thermodynamics is the study of energy transformation (or energy conversion) in a system.

Topic question

What is a system as it relates to thermodynamics? **A system is the total matter in a given area that transfers energy from one component to another.**

6.7 FIRST LAW OF THERMODYNAMICS

- The first Law of Thermodynamics states that energy can neither be created nor destroyed in a system.

Topic question

If energy is neither created nor destroyed in a system, where does the energy go when chemical potential energy is converted into chemical kinetic energy is an organism? **This may be hard for the student to answer, but it is important. When an organism converts chemical potential energy into chemical kinetic energy, there is work that is done. The kinetic energy allows the chemical reaction to occur. The chemical reaction is one of the forms of kinetic energy. This is only a small portion of the energy that is transformed into kinetic energy. The rest of the potential energy is converted into heat. Heat is another form of kinetic energy and is actually where most of the potential energy goes during energy conversions in living organisms. If the energy used to perform the reaction is added to the energy lost as heat, it would equal the starting potential energy and the first Law is proven true.**

6.8 SECOND LAW OF THERMODYNAMICS

- The Second Law of Thermodynamics states that systems favor the generation of increasing entropy (or disorder).

Topic question

Entropy is randomness. If entropy is favored in systems, why do all living organisms not slowly fall apart over time as they perform more and more chemical reactions? **This is also difficult for many students to grasp. When energy is transferred, such as in a chemical reaction, the creation of increasing randomness (entropy) is favored. The main form of entropy in living systems is heat. Heat is uncontrolled, random energy. However, there is a limit to how much energy is lost as heat. In addition, one of the main functions of all organism—what they expend a lot of their chemical energy doing—is maintaining order, directly counteracting the second Law. If organisms did not expend such a large amount of their chemical energy to counteract the formation of entropy, they would indeed not continue to live.**

6.9 ENZYMES

- Enzymes are protein molecules that control the chemical reactions of all living organisms.

Topic question

What is an enzyme? **There may be more than one answer for this depending on how far along the student is. Following this section, the answer is an enzyme is a large protein molecule which regulates (controls) the chemical reactions which occur in a cell.**

6.10 FREE ENERGY AND REACTION TYPES

- Free energy is the amount of energy available to perform work in a system.
- There are two types of reaction types—exergonic and endergonic.

Topic question

What is free energy? **Free energy is the amount of energy available to perform work in a system or a reaction.**

6.11 EXERGONIC REACTIONS

- Exergonic reactions are reactions in which the free energy is out of the system. Generally exergonic reactions occur spontaneously and are impossible to regulate.
- The products of an exergonic reaction possess less energy than the reactants.

Topic question

Why are exergonic reactions not used by organisms? **Since exergonic reactions occur spontaneously, they are almost impossible to regulate. Living organisms need to be able to strictly control which reactions occur and when. If exergonic reactions were used, the cell could not regulate them.**

6.12 ENDERGONIC REACTIONS

- Endergonic reactions require the input of energy before they will get started. The products of an exergonic reaction possess more energy than the reactants.
- Endergonic reactions are easy for the cell to regulate and are the reactions all organisms use for their metabolic needs.

Topic question

Why does the cell use endergonic reactions? **Because they require an input of energy to start, which allows the cell to tightly control which reactions occur and when.**

6.13 ACTIVATION ENERGY

- Activation energy is the energy required for a reaction to get started.
- Endergonic reactions have a high-activation energy.

Topic question

What is the activation energy? **The energy required to start a chemical reaction.**

6.14 ENZYMES SOLVE THE ACTIVATION ENERGY PROBLEM

- Enzymes are catalytic molecules that lower the activation energy of endergonic reactions so they can occur in biological systems.

Topic question

Why are enzymes important? **They lower the activation energy of endergonic reactions enough that the reactions can occur at safe temperatures in biological systems.**

6.15 MODELS OF ENZYME FUNCTION: LOCK AND KEY; INDUCED FIT

- The molecule to which an enzyme binds is called the substrate. The area where the enzyme binds the substrate is called the active site.
- The enzyme fits onto the active site like a key fits into a lock.

Topic question

What is the induced-fit theory of enzyme-substrate binding? **The enzyme fits onto the substrate like a key fits into a lock. Once the enzyme binds the substrate at the active site, the enzyme then changes its shape slightly. The change in shape allows for the enzyme and the substrate to interact better during the chemical reaction.**

6.16 ENZYME TYPES

- Catabolic enzyme catalyze catabolic reactions.
- Anabolic enzymes catalyze anabolic reactions.
- Bidirectional enzymes can catalyze both the anabolic and the catabolic reaction.

Topic questions

What type of enzyme catalyzes a condensation synthesis reaction? **If the question is too general, give them the hint of catabolic, anabolic, or bidirectional. The answer is anabolic because a condensation synthesis reaction is an anabolic reaction.**

You have discovered a new enzyme. You find that it can catalyze a condensation synthesis reaction and a hydrolysis reaction. What category of enzyme is this? **Bidirectional.**

6.17 COENZYMES AND COFACTORS

- Coenzymes are organic molecules that assist enzymes to function properly.
- Cofactors are inorganic molecules that assist enzymes to function properly.

Topic question

What is an inorganic molecule that assists an enzyme to function properly called? **A cofactor.**

6.18 ATP (ADENOSINE TRIPHOSPHATE)

- ATP is the molecule which provides the energy needed by the enzyme to get the endergonic reactions started.

Topic question

Why is ATP an important molecule? **It provides the needed activation energy for endergonic reactions.**

6.19 ENZYMES ARE ENERGY COUPLERS

- Enzymes couple the energy released when a phosphate bond of ATP is broken to the reaction which the enzyme catalyzes.

TOPIC QUESTION

What is meant when enzymes are described as energy couplers? **Enzymes are able to add water to a phosphate bond of ATP. This releases a lot of energy. Enzymes are able to then take that energy and link it to the endergonic reaction, which the enzyme catalyzes. In doing so, the enzyme couples the energy released from ATP to the reaction that the enzyme catalyzes so the reaction can proceed.**

6.20 ADVANCED

- Enzymes know when to turn on and turn off through feedback inhibition.

Topic question

How does feedback inhibition work? **When the product level from a reaction that an enzyme catalyzes gets too low in the cell, the enzyme is stimulated to "turn on." This results in the enzyme starting the reaction that makes that product. As the reaction produces more and more of the product, the level of the product builds up inside the cell. Once the level of the product gets high enough, the level stimulates the enzyme to "turn off," and the reaction stops until the product level gets low again.**

6.22 KEY CHAPTER POINTS

- Metabolism is the process of breaking down or building up organic molecules.
- All organisms convert potential energy stored in molecules into kinetic energy. This is done following the Laws of Thermodynamics.
- Exergonic reactions result in a net output of energy for the chemical reaction.

- Endergonic reactions result in a net input of energy for the chemical reaction. The energy required to get endergonic reactions going is called activation energy.

- The reactions of life are endergonic reactions and require a high activation energy. Enzymes are special proteins that lower the activation energy so endergonic reactions can occur safely in all cells.

- Only one enzyme can catalyze one reaction. Enzymes and the molecules they act on, called substrates, fit together like a lock and key.

- Enzymes that catalyze reactions to build molecules are called anabolic enzymes. Enzymes that catalyze reactions to break molecules down are called catabolic enzymes. Some enzymes can catalyze the anabolic and catabolic reaction and are called bidirectional enzymes.

- Enzymes are able to release the energy stored in ATP and couple that energy to lower the activation energy and fuel endergonic reactions.

- Enzymes know when to start and stop by feedback inhibition.

7 | Photosynthesis

7.0 CHAPTER PREVIEW

In this chapter we will:

- Describe the ways in which organisms obtain their energy and make their cell mass.
- Investigate the components and properties of the electromagnetic spectrum, including the visible light spectrum.
- Discuss the absorption spectrum of plant leaves and how that relates to photosynthesis.
- Study the plant structures and molecules that perform photosynthesis.
- Learn that although photosynthesis occurs in many linked biochemical reactions, it is a two-part process—a light-absorbing/energy-transforming event as well as a carbon fixation event.

7.1 OVERVIEW

- Carbon fixation is accomplished through photosynthesis.

Topic question

What is carbon fixation and how does it relate to photosynthesis? **Carbon fixation is the process of incorporating carbon into organic molecules. Photosynthesis is the set of biochemical reactions plants perform by using the sun's energy to fixate carbon into organic molecules.**

7.2 AUTOTROPHS

- An autotroph is an organism that can make its own cell mass and organic molecules.
- A photoautotroph is an organism that can make its own cell mass and organic molecules from carbon dioxide and the sun's energy.
- A chemoautotroph is an organism that can make its own cell mass and organic molecules using the energy obtained from inorganic molecules.

Topic question

What type of autotroph is a plant and why? **A plant is a photoautotroph because plants make their own cell mass and organic molecules using energy from the sun.**

7.3 HETEROTROPHS

- Heterotrophs cannot make their own cell mass from non-organic molecules as the autotrophs can. Heterotrophs make their cell mass and organic molecules by processing organic molecules, which they ingest.

Topic question

How are heterotrophs and autotrophs related? **Autotrophs bring carbon into the organic molecules so that both autotrophs and heterotrophs can use the organic molecules for energy. Also, the organic molecules that the autotrophs make are then used by heterotrophs to build their own organic molecules.**

7.4 ELECTROMAGNETIC SPECTRUM

- The electromagnetic spectrum is the total amount of radiant energy emitted by the sun.
- The electromagnetic spectrum includes energy with different wavelengths.

Topic questions

How does radiant energy travel? **In discreet units called photons.**

What are the components of the electromagnetic spectrum? **Radio waves, microwaves, ultraviolet waves, light waves, infrared waves, x-rays, and gamma rays.**

7.5 WAVELENGTH

- Wavelength is the distance from one peak of an electromagnetic wave to the next peak of the same wave.

Topic questions

What has a longer wavelength, infrared rays or microwaves? **Microwaves.**

The portion of the electromagnetic spectrum taken up by light represents the largest portion of the spectrum. True or False? **False. The portion of the spectrum taken up by light actually is the smallest portion of the electromagnetic spectrum.**

7.6 VISIBLE SPECTRUM

- White light is another name for the visible spectrum.

Topic question

What portion of the electromagnetic spectrum provides the energy for photosynthesis to occur? **Photosynthesis is driven by the energy from the visible spectrum (or white light).**

7.7 REFRACTION

- Refraction is the bending of light waves—each to a different degree, based on the wavelength of the light wave.

Topic question

When white light is passed through a prism, how and why does the light coming out of the prism differ from the light entering the prism? **A prism has the ability to refract light. Light entering a prism appears white. When it passes through a prism, the components of the white light are refracted (or bent) at different angles because they have different wavelengths. Light exiting the prism has been bent to different degrees so that all of the components of the white light have been broken up. Light emerging from a prism looks like a rainbow and has the main colors of red, orange, yellow, green, blue, indigo, and violet.**

7.8 REFLECTION

- The color of an object depends on the wavelength of light it reflects.

Topic questions

Why does a blue balloon appear blue? **When white light strikes an object, some of the wavelengths of light are absorbed and some are reflected. The wavelengths that are absorbed cannot be seen. For example, a blue balloon absorbs all wavelengths of white light except blue. The blue wavelength of light is reflected to our eye. When the light with the blue wavelength strikes the eye, it is perceived as blue. Therefore, a blue balloon appears blue because it reflects light in the blue wavelength.**

What is the absorption spectrum of chlorophyll and why is it important to understand? **The absorption spectrum of chlorophyll includes all the wavelengths of light that chlorophyll absorbs. The absorption spectrum of chlorophyll includes all colors—except green and some yellow. (This means that chlorophyll reflects green and some yellow wavelengths of light, which is why leaves look green). It is important to know this because the energy provided by the sun to fuel photosynthesis comes from the light chlorophyll absorbs.**

7.9 PHOTOSYNTHESIS: TWO SEPARATE, BUT LINKED, REACTIONS

- Photosynthesis occurs as two sets of separate, but linked, chemical reactions occurring in the chloroplasts of the green parts of plants.

Topic question

In which organelle does photosynthesis occur? **Chloroplasts.**

7.10 THYLAKOID

- The thylakoid is a folded membrane system inside of chloroplasts.
- The light-dependent reactions occur in the thylakoid membranes.

Topic question

What are grana? **They are the flattened and stacked discs of the thylakoid.**

7.11 CHLOROPHYLL

- Chlorophyll is a type of molecule called a pigment. Chlorophyll has the ability to absorb energy from the sun.
- Chlorophyll is contained in the thylakoid.

Topic questions

What are the two types of chlorophyll in most plants? **Chlorophyll a and b.**

What does chlorophyll "do"? **Chlorophyll absorbs all wavelengths of light except for green and some yellow wavelengths. (This absorbed energy is used by more chemical reactions to make glucose from carbon dioxide and water).**

7.12 CAROTENOIDS

- Carotenoids are other types of photosynthetic pigments.

Topic question

Why are carotenoids important? **They enhance the absorption spectrum of chlorophyll so that more of the sun's energy can be used in photosynthesis.**

7.13 FALL COLORS

- The fall colors are due to the presence of carotenoids in the leaves.

Topic question

Why do leaves change colors in the fall? **Because the plants stop producing chlorophyll during the fall. As the level of the chlorophyll drops in the leaves, they lose their green color. They begin to appear as other colors because the carotenoids are still present in the leaves. Since the carotenoids absorb green light and reflect other wavelengths (mainly red, orange, yellow, and brown, depending on the type of carotenoid) the leaves take on the color of the carotenoid that is present.**

7.14 PHOTOSYNTHESIS

- Photosynthesis is broken into the light-dependent reactions and the chemical reactions of the Calvin cycle.
- The first step of photosynthesis are the light-dependent reactions. During this series of chemical reactions, energy from the sun is absorbed and transferred to other molecules.
- The second series of reactions is the Calvin cycle. During the Calvin cycle, the energy that was transferred to the molecules in the light-dependent reactions is transferred again to enzymes in the Calvin cycle. During the transfer of energy, glucose is made from carbon dioxide and water molecules.

Topic question

What is the overall process of photosynthesis? **The light-dependent reactions absorb energy from the sun and transfer it to energy-absorbing molecules. This energy is then transferred to enzymes in the Calvin cycle to fuel the endergonic reactions, which make glucose from carbon dioxide and water.**

7.15 LIGHT-DEPENDENT REACTIONS

- When the sun's energy is absorbed by the pigments of the light reactions high energy electrons are released from chlorophyll. Also, a water molecule is split apart. This releases electrons to replace those lost by chlorophyll. The high energy electrons provide the energy needed for the Calvin cycle.

Topic question

What is the overall purpose of the light reactions? **To release high-energy electrons from chlorophyll and capture their energy. This energy is then harnessed and used later in the Calvin cycle.**

7.16 PHOTOSYSTEMS I AND II

- Photosystem I and photosystem II are responsible for generating and passing along high energy electrons.

Topic questions

Where are the enzymes of the light reactions contained? **The membranes of the thylakoid.**

Which photosystem is activated first? **Photosystem II.**

Where are activated electrons obtained? **From chlorophyll, which releases electrons. The energy of the sun is captured by the enzymes of photosystem I and photosystem II and passed to the electrons.**

7.17 LIGHT REACTIONS MAKE ATP (ADENOSINE TRIPHOSPHATE) FOR USE IN THE CALVIN CYCLE

- The electron transport chain takes the activated electrons from PS II to PS I; in the process, ATP is made.

Topic questions

How is ATP made in the electron transport chain? **As the electron is passed through the electron transport chain, it loses energy. At key points, the lost energy is harnessed by an enzyme in the electron transport chain and used to fuel the endergonic reaction, which makes ATP from ADP.**

Where is this ATP needed? **This ATP is then used for energy during the Calvin cycle.**

7.18 LIGHT REACTIONS MAKE NADPH FOR USE IN THE CALVIN CYCLE

- Electrons are excited a second time in PS I. The activated electron is accepted by NADP, and the energy from the electron is transferred to the molecule NADP, making NADPH.

Topic questions

How is NADPH generated? **The activated electron from PSI is transferred to NADP, forming NADPH. This electron is in the form of a hydrogen ion (H), which is why the NADP turns into HADPH when the electron is accepted.**

What does NADPH do? (What is its function?) **NADPH transfers the energy of the high-energy electron from the light reaction to the Calvin cycle. The energy is needed for the endergonic reactions of the Calvin cycle.**

7.19 THE CALVIN CYCLE

- The Calvin cycle can occur with or without light.
- During the Calvin cycle, energy from the ATP and NADPH generated in the light reactions is used to make a three-carbon sugar.

Topic questions

Where are the atoms for the carbohydrates made in the Calvin cycle obtained? **The carbon is obtained from carbon dioxide. Hydrogen and oxygen are obtained from water molecules.**

Where does the energy come from to fuel the endergonic reactions which the Calvin cycle uses to make the three-carbon sugar? **The energy comes from the ATP and the NADPH, which was made during the light-dependent reactions.**

7.20 PRODUCTION OF G-3-P

- The ultimate product of the Calvin cycle is the three-carbon sugar glyceraldehyde-3-phosphate (G-3-P). Usually two G-3-P molecules are linked together to form one six-carbon glucose molecule.

Topic question

What does the plant usually do with the G3P molecules? **Link two of them together to form one six-carbon glucose molecule.**

7.21 ADVANCED

- There are two pathways the plant can use for "photosynthesis." They are called the cyclic and non-cyclic pathways. The pathway we have learned about this entire chapter is the most common and is the non-cyclic pathway.

Topic questions

What does the "cyclic" and "non-cyclic" pathway refer to? **The cyclic and non-cyclic refer to the pathway the electrons take once they are excited by the sun.**

What is the movement of electrons in the cyclic pathway? **The cyclic pathway begins when the sun activates electrons from PS I. The electrons come from chlorophyll. The electrons become activated, then move into the electron transport chain. ATP is made when the electrons fall through the chain. Then the electrons return to the chlorophyll from where they came.**

7.23 KEY CHAPTER POINTS

- Photosynthesis is a set of chemical reactions photoautotrophs use to convert the sun's energy into carbohydrate molecules.

- Autotrophs are the initial source of all carbon containing molecules on earth.

- The process of bringing carbon into organic molecules is called carbon fixation.

- The sun emits energy in many different forms; this is called the electromagnetic spectrum.

- Plants use energy from the visible light portion of the electromagnetic spectrum for photosynthesis. Only the light absorbed by plants can be used for photosynthesis.

- Photosynthesis occurs in the specialized structures called chloroplasts, which contain the light-absorbing pigments chlorophyll and carotenoids.

- Photosynthesis is two separate—but linked—sets of chemical reactions. The light-dependent reactions absorb the sun's energy, and the Calvin cycle uses that absorbed energy to synthesize glucose.

- Glucose is then made available to the plant and heterotrophs for use in generating molecules of ATP and synthesizing other organic molecules.

- There are two photosystems in plants that act together to excite and pass on high-energy electrons.

- The non-cyclical pathway utilizes PS I and II to generate activated electrons and make ATP and NADPH for use in the Calvin cycle. Electrons do not return to the same molecules from which they came.

8 Cellular Respiration

8.0 CHAPTER PREVIEW

In this chapter we will:

- Investigate the structure and function of ATP.
- Learn how ATP is utilized by enzymes to fuel endergonic reactions they catalyze.
- Investigate the biological processes of aerobic and anaerobic respiration.
- Study the internal structure of mitochondria and how it relates to aerobic respiration.

8.1 OVERVIEW

- ATP is used by almost all organisms for energy.
- ATP is made from the energy released from the metabolism of glucose.

Topic question

Why is ATP made from glucose? **Glucose cannot be used directly by organisms for energy. Instead, the glucose is "burned" and ATP is made for the energy released from the bonds of the glucose.**

8.2 THE LINK BETWEEN GLUCOSE, ATP, AND CELLULAR RESPIRATION

- Glucose cannot be used as a direct energy source by organisms.
- ATP must be used to fuel the endergonic reactions all organisms perform.
- The process of making ATP molecules by burning glucose molecules is called cellular respiration.

Topic question

What is cellular respiration? **The process of breaking the bonds of glucose molecules and harnessing the energy released to make ATP.**

8.3 ATP IS USED TO FUEL THE ENDERGONIC REACTIONS OF LIFE

- ATP is used as the direct energy source for the endergonic reactions of all organisms.
- Energy is released from ATP when water is added to a phosphate bond. Adding water to a bond and breaking it is called hydrolysis.

Topic questions

What is ATP used for? **It provides the direct energy needed for the cell to perform the endergonic reactions of life.**

The enzymes that hydrolyze ATP are under virtually no regulation and hydrolyze the phosphate bonds of ATP all the time. True or False? **False. These enzymes are strictly regulated.**

What is ATPase? **ATPase is the generic name for an enzyme that can add water to a phosphate bond of ATP and hydrolyze (break) it to release energy.**

8.4 AEROBIC AND ANAEROBIC RESPIRATION

- Aerobic respiration is the metabolism of glucose using oxygen to generate ATP.
- Anaerobic respiration is the metabolism of glucose without oxygen to make ATP.

Topic questions

What is the benefit of anaerobic respiration? **It generates ATP much more quickly than aerobic respiration.**

What is the benefit of aerobic respiration? **It generates much more ATP per molecule of glucose metabolized than anaerobic respiration does. This means it is more efficient.**

8.5 THE CHEMICAL EQUATION FOR CELLULAR RESPIRATION

- Cellular respiration is a series of chemical reactions that occur in a step-wise fashion.
- During cellular respiration, glucose is metabolized into carbon dioxide and energy is released. This energy is used to make ATP.

Topic question

What happens to the carbon atoms in glucose during cellular respiration? **They are metabolized (broken down) into carbon dioxide.**

8.6 FOUR STEPS OF AEROBIC CELLULAR RESPIRATION

- There are four main steps to aerobic respiration—glycolysis, the transition reaction, the Krebs cycle, and the electron transport chain. Each one is a set of complex chemical reactions.

Topic question

What are the four steps of aerobic respiration? **Glycolysis, the transition reaction, the Krebs cycle, and the electron transport chain.**

8.7 MITOCHONDRIA: STRUCTURE AND FUNCTION

- All the steps of aerobic respiration occur in mitochondria except glycolysis.
- Mitochondria have a complex structure that accommodates the needs of aerobic respiration.

Topic questions

How many membranes do mitochondria have? **Two lipid bilayers. They are a double membrane-bound organelle. If the student answers four membranes, that is also technically true since there are two lipid bilayers.**

What are the membranes called? **The inner and outer membranes.**

What are cristae? **Folds of the inner membrane.**

8.8 GLYCOLYSIS

- Glycolysis occurs in the cytoplasm, outside of the mitochondrion.
- Although glycolysis is not considered part of aerobic respiration, aerobic respiration cannot proceed without the pyruvate that is produced during glycolysis.
- Aerobic respiration occurs entirely inside of mitochondria.

Topic question

Where does glycolysis occur? **Outside of the mitochondrion.**

What happens during glycolysis? **Glucose is converted into two three-carbon molecules called pyruvate. In addition, two molecules of ATP and two molecules of NADH are made.**

Where does aerobic respiration occur? **Inside the mitochondria.**

8.9 TRANSITION REACTION

- When the pyruvate first enters the mitochondria, it undergoes a transition reaction.
- The transition reaction results in the production of one molecule of carbon dioxide and one molecule of acetyl coenzyme A per molecule of pyruvate that undergoes the reaction.

Topic question

What are all the products of the transition reaction? **Pyruvate is metabolized into one molecule of carbon dioxide and one molecule of acetyl coenzyme A. Also, two molecules of NADH are produced.**

8.10 MOLECULAR ACCOUNTING THROUGH THE TRANSITION REACTION

- Each six-carbon glucose molecule is metabolized into two three-carbon pyruvate molecules (glycolysis). Then each pyruvate molecule is metabolized into a molecule of acetyl coenzyme A and a carbon dioxide molecule (transition reaction).
- Two molecules of ATP are made during glycolysis and two molecules of NADH are made during glycolysis and a total of two molecules of NADH are made during the transition reaction (one per molecule of pyruvate).

Topic question

How many molecules of ATP are produced per glucose molecule during glycolysis? **Two.**

How many molecules of NADH are produced per glucose molecule during the transition reaction? **Two.**

8.11 THE KREBS CYCLE

- Acetyl coenzyme A then enters into the Krebs cycle.
- The Krebs cycle takes place in the mitochondrial matrix.
- The Krebs cycle is also called the tricarboxylic-acid cycle and the citric-acid cycle because these are all key molecules formed during the cycle.
- The Krebs cycle is a series of reactions resulting in the remaining carbon molecules from glucose getting metabolized into carbon dioxide while producing ATP, NADH, and $FADH_2$.
- NAD and FAD are coenzymes used to transport activated electrons to the electron transport chain. The activated electrons are in the form of hydrogen ions.

Topic questions

Where does the Krebs cycle occur? **The mitochondrial matrix.**

What is the "starting" molecule for the Krebs cycle? **Acetyl coenzyme A.**

What is produced during the Krebs cycle? **ATP, NADH, $FADH_2$, and carbon dioxide.**

What are NAD and FAD converted to when they capture an activated electron? **NADH and $FADH_2$, respectively.**

8.12 TURNS OF THE KREBS CYCLE AND ENERGY

- The Krebs cycle "turns" twice for every molecule of glucose, which starts the cycle.
- For each turn of the Krebs cycle, one molecule of ATP, three molecules of NADH and one molecule of $FADH_2$ are produced.

Topic question

How many molecules of ATP, NADH, and $FADH_2$ are produced in the Krebs cycle per molecule of glucose? **Two molecules of ATP, six molecules of NADH and two molecules of $FADH_2$.**

8.13 TURNS OF THE KREBS CYCLE AND CO_2

- Two molecules of carbon dioxide are produced for every turn of the Krebs cycle.

Topic questions

How many molecules of carbon dioxide are produced during the Krebs cycle per molecule of glucose? **Four.**

How many molecules of carbon dioxide total are produced during the aerobic metabolism of glucose? **Six.**

8.14 THE ELECTRON TRANSPORT CHAIN

- The electron transport chain takes place in the cristae.
- The electron transport chain generates thirty-two molecules of ATP.

Topic questions

Where does the energy come from to make ATP in the electron transport chain? **The energy is obtained from the activated electrons of NADH and FADH$_2$. These molecules were made during glycolysis, the transition reaction, and the Krebs cycle.**

Why do organisms need oxygen for aerobic respiration? **Because oxygen serves as the final electron acceptor for the electron transport chain. The electrons are in the form of hydrogen ions; when they combine with oxygen, water is formed.**

8.15 ENERGY CONVERSION AND AEROBIC RESPIRATION

- Only about 40% of the available energy in the bonds of glucose is transformed into energy contained in ATP. The rest is lost as heat.
- The ATP then moves out into the cytoplasm so it can be used by enzymes to fuel endergonic reactions.

Topic question

What happens to the ATP once it is made? **It can either stay in the mitochondria to be used as energy there or pass out of the mitochondria and into the cytoplasm. Once in the cytoplasm, ATP is available to be used by whatever enzyme may need it.**

8.16 ANAEROBIC RESPIRATION

- Anaerobic respiration is also called fermentation.
- Anaerobic respiration does not need oxygen.
- Glycolysis is the first step of anaerobic respiration.
- Anaerobic respiration does produce ATP quickly, but in lesser amounts than aerobic respiration does.

Topic questions

How many molecules of ATP are made during anaerobic respiration of glucose? **Two.**

What is lactic acid? **It is the end product of fermentation (anaerobic respiration) in animal cells.**

8.17 ADVANCED

- The proton motive force is the ability of hydrogen to perform work as a result of a concentration gradient. Chemiosmosis is a result of the proton motive force. It is hydrogen ions flowing across the cristae membranes and causing ATP synthase to make ATP from ADP.

Topic questions

How does the proton motive force develop? **The proton motive force is the ability of hydrogen ions to perform work. It develops because proton pumps actively pump hydrogen ions across the cristae membrane. The hydrogen ions build up in high level on one side of the cristae. As the protons (hydrogen ions) move back across the membrane as a result of gated diffusion, their movement across the membrane through the protein gate drives the enzyme ATP synthase. This makes ATP from ADP.**

What is chemiosmosis? **It is the process of the hydrogen ions flowing across the gated channel providing the energy for ATP synthase to work.**

In which two organelles do you know that the proton motive force and chemiosmosis is responsible for making ATP? **Mitochondria and chloroplasts.**

8.19 KEY CHAPTER POINTS

- Photosynthesis generates molecules of glucose. Glucose cannot be used by organisms as a direct energy source.

- Glucose is processed further by all organisms through the chemical reactions of cellular respiration. Cellular respiration extracts the energy contained in glucose and uses it to make molecules of ATP.

- ATP is the energy source organisms use to fuel the endergonic reactions of life.

- Enzymes hydrolyze the phosphate bonds of ATP, which releases energy. The enzymes then couple that energy to fuel the reactions they catalyze.

- Cellular respiration can be performed using oxygen or not using oxygen, depending on the conditions of the cell. Cellular respiration using oxygen is called aerobic respiration. Cellular respiration not using oxygen is called anaerobic respiration.

- Aerobic respiration, like photosynthesis, is a set of related chemical reactions. There are three steps—the first occurs in the cytoplasm while the second and third occur in the mitochondria.

- The first step of cellular respiration, glycolysis, occurs in the cytoplasm and is technically not considered aerobic respiration. It is not considered part of cellular respiration because glycolysis is performed by the cell at other times and for reasons other than aerobic respiration.

- The next two steps—the Krebs cycle and the electron transport chain—occur in the mitochondria. Both are aerobic respiration.

- Aerobic respiration generates thirty-six molecules of usable ATP per molecule of glucose metabolized.

- The structure of mitochondria is specialized to perform aerobic respiration efficiently.

- Mitochondria use the proton motive force to drive the enzymes that make ATP, called ATP-ase.

- Oxygen is consumed and carbon dioxide is produced during aerobic respiration.

- Anaerobic respiration occurs in the cytoplasm and generates two molecules of usable ATP per molecule of glucose metabolized.

9 DNA, RNA, and Proteins

9.0 CHAPTER PREVIEW

In this chapter we will:

- Investigate the structure and function of DNA and RNA.
- Discuss the organization of DNA into chromosomes and genes.
- Study the concept that one gene codes for the production of one protein.
- Introduce the concept that genes control all traits of an organism or cell.
- Explore the biological processes of eukaryotic transcription and translation.
- Discuss transcription and translation in prokaryotes.
- Learn what the genetic code is and become comfortable using it.

9.1 OVERVIEW

- DNA contains the genetic information.
- Every trait of an organism is controlled by DNA.

Topic question

How does DNA communicate with the rest of the cell? **Through polypeptides. DNA contains the information that instructs the cell as to which proteins to make and when those proteins should be made. The message from the DNA is made into an RNA molecule. This message is the instruction for the cell to make a specific polypeptide (protein). The RNA molecule then is read by a ribosome in the cytoplasm and the protein is synthesized. The protein the cell makes causes the cell to behave in a certain way.**

9.2 DNA GENERAL

- DNA is tightly packaged around protein molecules called histones.
- DNA is made from only four different nucleotides—adenine, thymine, cytosine, and guanine.

Topic questions

What is the basic structure of DNA? **A central five-carbon sugar (called a pentose), which is deoxyribose. One end of the deoxyribose has a phosphate group attached to it, the other end has one of four nitrogen containing bases.**

What are the four nitrogenous bases of DNA? **Adenine (A), thymine (T), guanine (G), and cytosine (C).**

How are the nucleotides of DNA named? **They are named after the nitrogenous base contained in the nucleotide.**

9.3 DNA IS MADE OF PURINES AND PYRIMIDINES

- All nucleotides share a common structural design.
- C and T are pyrimidines. A and G are purines.

Topic questions

What is the main difference between the purines and the pyrimidines as far as molecular structure is concerned? **The purines have a double-ring structure in their nitrogen base and the pyrimidines have a single-ring structure.**

What are the three components of a nucleotide? **A central five-ring sugar, a phosphate group, and a nitrogenous base.**

9.4 DNA IS A DOUBLE-STRANDED HELIX

- The structure of DNA is a double-stranded helix.
- Each strand of DNA is called a complementary strand.
- Only T and A can base pair with one another, and only C and G can base pair with one another.
- Chargaff's rule is that the percentage of thymine and adenine are equal in DNA, as is the percentage of C and G.

Topic questions

What is base pairing? **Base pairing is another way of stating that the nucleotides of DNA form hydrogen bonds with one another.**

Can C and T base pair? **No, only A and T can pair with one another and only C and G can pair with one another.**

What does "complementary strands" mean? **This is often a hard concept for students to understand; there may be several ways they can answer which are all correct. DNA exists as two complementary strands. One strand is complementary to the other because they form hydrogen bonds, which cause the two strands to stick together. Or, they are complementary because one strand is sequenced so that every time there is an A in the strand, there is a T on the strand directly across from it, allowing A and T to base pair. Likewise, every time there is a C on one strand, there is a G on the other strand and they base pair. The nucleotides on one strand always being able to base pair with the nucleotides on the other strand causes the two DNA strands to bond to one another and form the double-stranded molecule.**

9.5 DNA ORGANIZATION

- Eukaryotic DNA is broken into smaller segments called chromosomes.
- The units of DNA can appear as compact structures called chromosomes or as loosely coiled structures called chromatin.

Topic questions

What are the smaller units into which eukaryotic DNA is broken? **Chromosomes.**

What is the difference between chromatin and chromosomes? **Chromatin is the loosely packed form of chromosomes.**

9.6 CHROMOSOMES, GENES, AND TRAITS

- The information in chromosomes is broken into small segments called genes.
- One gene contains the information to make one protein.
- The genetic code is the orderly nucleotide arrangement of DNA that results in the coding of information on how to make a protein.
- Genes control the traits of an organism.

Topic questions

What is a gene? **It is the individual functional unit of the chromosome. One gene contains the information for (or "codes for") the production of one protein.**

What is the genetic code? **The orderly arrangement of nucleotides of a gene that contain the information for making a protein.**

9.7 RNA

- RNA is ribonucleic acid. It is single-stranded.

Topic question

What are the four nucleotides of RNA? **Adenine, uracil, cytosine, and guanine.**

9.8 mRNA

- mRNA carries the message from the DNA to the ribosomes regarding which protein is to be made.

Topic question

Where does an mRNA molecule start and where does it end up? **The mRNA molecule is made in the nucleus. It then travels out of the nucleus into the cytoplasm, where it binds to one or more ribosomes.**

9.9 tRNA

- tRNA brings the needed amino acids to the ribosome during protein synthesis.

Topic question

How is the cross shape of tRNA maintained? **Through hydrogen bonding between the nucleotides of the tRNA molecule.**

9.10 FROM DNA TO PROTEIN

- The sequence of production of a protein starting from the DNA is:
  - the sequence of gene nucleotides is made into a complementary sequence of RNA nucleotides (this is mRNA) in the nucleus
  - mRNA is processed in the nucleus
  - the mRNA moves from the nucleus into the cytoplasm
  - one or more ribosomes bind to the mRNA
  - the sequence of nucleotides in the mRNA is read by the ribosomes
  - tRNA brings amino acids to the ribosomes for protein synthesis
  - the protein is synthesized by ribosomes in the cytoplasm.
- The linear sequence of nucleotides in DNA serves as the code for protein synthesis.

Topic question

What is the relationship between a gene (DNA) and the protein for which it codes? **The sequence of DNA.**

9.11 CODONS

- A codon is a group of three nucleotides in a gene that codes for one specific amino acid.
- There are three types of codons—ones coding for an amino acid, start codons, and stop codons.

Topic questions

What is a codon? **It is a group of three nucleotides that code for the insertion of a specific amino acid in a specific location in a protein.**

What is a start codon? **It is a group of three nucleotides positioned where the ribosomes always start to read an mRNA molecule and begin protein synthesis.**

Why are stop codons stop codons? **Stop codons are stop codons because they do not code for any amino acid. Therefore, when the ribosome reaches a stop codon, protein synthesis stops because there is no amino acid to insert into the protein.**

9.12 READING FRAMES AND CODONS

- DNA and mRNA are read by enzymes linearly. That is, they are read as you would read a sentence in a book. This is called the reading frame.

Topic question

Why is the reading frame concept important? **It is important to understand that the DNA and RNA molecules are read in a straight line because this is the only way the proper** protein can be made. If codons or nucleotides are skipped, or if segments of a gene or mRNA are read twice, then a protein will be synthesized that does not have the proper sequence. Kids sometimes have a hard time with this. However, the reading frame of DNA and RNA is exactly the same concept as the reading frame we use to read a book. We read in a linear fashion. We start at the top left of the book and read from left to right. That is the only way the information makes any sense because that is how it was designed to be read. DNA and RNA are the same way. If the reading frame is not maintained, it would be like reading a book, and every three words going back two words and then starting to read again. Or, it would be like trying to read every fourth letter in a sentence. Neither of those reading methods would make any sense because the reading frame is not maintained. If a person were to read like that, the sentence would not make any sense. If DNA were read like that, it would lead to the production of an incorrect protein structure.

9.13 TRANSCRIPTION

- Transcription is making mRNA from DNA.

Topic questions

Describe how transcription works. **Helicase unwinds the segment of DNA containing the gene to be transcribed into mRNA. RNA polymerase binds to the DNA and starts making mRNA. The RNA polymerase moves along the gene. As it does, it forms the mRNA molecule by base pairing mRNA nucleotides across from the DNA nucleotide template. When the RNA polymerase reaches the end of the gene, it stops making mRNA.**

What happens when RNA polymerase reaches an A in the DNA? **It inserts a U into the mRNA molecule because RNA does not use thymine, it uses uracil to base pair with adenine.**

9.14 DNA CODONS ARE MAINTAINED IN mRNA

- Since mRNA is made directly from the DNA template, the codon sequence of DNA is maintained in the codon sequence of mRNA.

Topic question

Why is the concept that mRNA is made directly from a DNA template important? **Since the mRNA is made from the DNA, the codon sequence of the DNA is maintained in the mRNA. Also, the reading frame is maintained. This ensures the proper amino acid sequence in the protein.**

9.15 mRNA PROCESSING OCCURS AFTER TRANSCRIPTION

- In eukaryotes, once the mRNA molecule is made, introns are removed from the molecules and the exons are spliced back together.
- Once the mRNA is processed, it leaves the nucleus and enters the cytoplasm.

Topic question

What is mRNA processing? **It is the process of removing pieces of the mRNA that are not needed for protein synthesis. These pieces are called introns. The introns are removed. The pieces left over are the exons. The exons are sliced back together, which forms the completed mRNA molecule.**

9.16 TRANSLATION

- Translation is the process of a ribosome making a protein from an mRNA template.

Topic question

How is translation performed? **The ribosome starts reading the mRNA at the start codon. It then moves along the mRNA reading the codons one at a time, assembling the protein.**

9.17 THE RIBOSOME PAIRS THE mRNA CODON TO A tRNA ANTICODON

- tRNA brings the amino acids to the ribosomes.
- The ribosome links the amino acids together by matching the codon of the mRNA to the anticodon of the tRNA.

Topic questions

How does the ribosome know that a particular tRNA molecule is carrying the proper amino acid the ribosome needs? **The tRNA has a section on it called the anticodon. The anticodon needs to be complimentary to the codon in order for the ribosome to know the tRNA is carrying the proper amino acid. If the anticodon is complimentary to the codon, they form hydrogen bonds (they base pair), which allows the ribosome to link the amino acid to the growing protein.**

When a protein is being synthesized by a ribosome, how many tRNA molecules are normally attached to the mRNA at one time? **Normally, two tRNA molecules are attached to the mRNA for each ribosome at any one time. As the ribosome moves down the mRNA, one tRNA falls off as another base pairs to the codon.**

9.18 PROTEINS ARE PROCESSED AFTER THEY ARE TRANSCRIBED

- Signal sequences are added to proteins in the ER after they are synthesized by the ribosomes.

Topic question

What is a signal sequence? **It is a small number of nucleotides that are added to a protein in the ER and tell the cell where the protein is supposed to be taken after it is made.**

9.19 GENE EXPRESSION

- When a gene is expressed, it means that the protein the gene codes for is being made.
- Another way of stating it is that when a gene is expressed, the trait the gene codes for is visible in some way.

Topic question

What is meant by saying that a gene is being expressed? **Either the protein for which the gene codes is being synthesized by the cell or the trait the gene codes for is visible in some way.**

9.20 THE PRACTICAL GENETIC CODE

- The table listing all the mRNA codons is presented for the students to learn how to use.

Topic questions

Using the codon table, what are the mRNA codons for threonine (thr)? **ACU, ACC, ACA, ACG.** What are the DNA codons for the same amino acid? **TGA, TGG, TGT, TGC.**

If an mRNA codon contains the following sequence—AUU UCU GGU CGC CAA UGA—what is the sequence of amino acids in the protein? **Isoleucine (Ilu) Serine (Ser) Glycine (Gly) Arginine (Arg) Glutamine (Gln) Stop**

9.22 KEY CHAPTER POINTS

- DNA is the nucleic acid that contains the genetic code.
- DNA communicates to the rest of the cell and organism using proteins.
- The highly-ordered nucleotides of DNA determine which proteins are made and when.
- DNA is made up of four nucleotides—adenine, thymine, cytosine, and guanine. They link together in such a way that DNA forms a double-stranded helix.
- Each strand of DNA is complementary to the other.
- Genes are the smallest unit of the genetic code. Each gene codes for the production of one protein.
- DNA replicates itself using a variety of enzymes, including DNA polymerase, prior to the cell dividing.
- The code of DNA is contained in three-nucleotide sequences called codons. Each codon codes for one amino acid in a protein.

- DNA codons are transcribed into mRNA codons using a variety of enzymes, including RNA polymerase. mRNA then exits the nucleus and goes into the cytoplasm.

- mRNA is decoded by ribosomes, and the protein encoded by the mRNA is translated into a protein by ribosomes and tRNA.

- Following translation, proteins are processed further by the cell.

10 | Cell Reproductions: Mitosis

10.0 CHAPTER PREVIEW

In this chapter we will:

- Discuss the biological process of asexual cell division, called mitosis.
- Review the theory of spontaneous generation and how it was disproved.
- Review the Theory of Biogenesis.
- Investigate the process of DNA replication.
- Learn how DNA is packaged in the cell at various stages.
- Discuss the events of the cell cycle.

Topic question

10.1 OVERVIEW OF CELL DIVISION

- Cell division, also called cell reproduction, is the process of one cell dividing into two cells.
- Mitosis, meiosis, budding, and binary fission are all types of cell division.

Topic questions

What is cell division? **The process of one cell dividing into two cells.**

How do multicellular organisms make more cells? **Mitosis.**

How do unicellular organisms make more cells? **Binary fission.**

10.2 ONE PARENT DIVIDES INTO TWO DAUGHTER CELLS

- The cell that divides is called the parent cell; the two cells formed are called the daughter cells.
- During cell division, a complete set of DNA is passed from the parent cell to each daughter cell.

Topic questions

What is important to understand about the passage of DNA from parent to daughter cell and why? **It is critical to understand that each daughter cell receives a complete copy of DNA from the parent cell. If the daughter cells do not receive a complete copy of DNA, they will not function properly.**

10.3 SPONTANEOUS GENERATION

- Spontaneous generation is the incorrect belief that living things can arise from nonliving things.
- Spontaneous generation was thoroughly disproved by Louis Pasteur in the mid 1800s.

Topic question

Describe the experiment by Pasteur that proved spontaneous generation was not true. **He designed flasks with a gooseneck in them. In this way, he could fill the flask with broth that would provide favorable conditions for the growth of the spores and leave the flask exposed to the air. However, he thought the gooseneck would trap any microscopic organisms and prevent them from ever reaching the broth. He boiled the broth; this killed any organisms (and their spores) in the broth and the neck of the flask at the beginning of the experiment. After boiling the broth, he waited for a week, and there was no growth of microbes in the broth. He then tipped one of the flasks to** let the broth touch the bend, thinking this would contaminate it with all the microbes that had been trapped there. Soon after that, microbes grew in the broth. But there was never any growth in the boiled flasks that were not tipped. By allowing air to get into and out of the flask, he proved that spontaneous generation does not occur. If it did, there would have been growth in the broth regardless of the presence of the gooseneck.

10.4 THE THEORY OF BIOGENESIS

- The theory of biogenesis states that life can only come from life.

Topic question

What is the theory of biogenesis? **The theory that life can only come from other living things.**

10.5 CELL DIVISION—BASICS

- There are two basic types of cell reproduction—sexual and asexual.

- Mitosis is a type of asexual cell division/reproduction.

- The purpose of mitosis is to ensure that each daughter cell receives a complete copy of DNA from the parent cell.

Topic questions

What type of reproduction is mitosis—sexual or asexual? **Asexual.**

What is the purpose of mitosis? **To ensure that each daughter cell receives a complete copy of DNA from the parent cell.**

Are cells formed from mitosis genetically identical or different? **Identical.**

What is the overall process of mitosis? **A parent cell copies its DNA (or chromosomes). As the parent cell splits into two daughter cells, one complete copy of DNA (or chromosomes) moves into one daughter cell and the other copy into the other daughter cell. When completed, mitosis results in the generation of two daughter cells from one parent cell. Each of the daughter cells has a copy of fully functional DNA.**

10.6 CELL LIFE CYCLE

- The cell cycle is the series of events occurring in a eukaryotic cell from one cell division to the next.

- The cell cycle is divided into interphase and mitosis.

Topic questions

What is the cell cycle? **The cell cycle is the cycle of events that occur in a eukaryotic cell from one cell division to the next.**

What are the phases of mitosis? **Prophase, prometaphase, metaphase, anaphase, and telophase.**

10.7 DNA REPLICATION

- The first step of cell division is DNA replication, which occurs at the end of interphase.

- Helicase is an enzyme that "unzips" DNA by breaking the hydrogen between complementary base pairs of DNA.

- DNA polymerase is the enzyme that links nucleotides together, forming a new daughter strand of DNA across from the parent strand template.

Topic questions

When does DNA replication occur in eukaryotic asexual cell division? **At the end of interphase.**

When are genes and chromosomes replicated? **Since chromosomes are simply smaller pieces of DNA that contain the genes, they are both replicated at the end of interphase. Often, the words DNA, genes, and chromosomes are used interchangeably.**

What is the function of helicase and DNA polymerase? **Helicase breaks the hydrogen bonds between the complementary DNA strands to unzip the DNA. DNA polymerase then can make a new DNA strand once the DNA is unzipped.**

10.8 CHROMOSOME TERMINOLOGY DURING CELL REPRODUCTION

- Once replicated, identical chromosomes remain attached to one another at the centromere.
- A chromatid is the name of one of the identical strands of a replicated chromosome.
- Sister chromatids are identical chromosomes that are attached to one another at the centromere.

Topic questions

What is a sister chromatid? **An identical copy of a chromosome still attached to its identical copy.**

Where are sister chromatids attached to one another? **At the centromere.**

10.9 INTERPHASE

- Interphase is the phase of the cell cycle that starts immediately after the cell divides.
- During the last part of interphase, the cell duplicates all its chromosomes.

Topic question

What occurs during interphase? **The cell performs all the functions for which it was designed. The last part of interphase is when the cell duplicates its chromosomes.**

10.10 PROPHASE

- Prophase is the first stage of mitosis.

Topic questions

If you are standing inside a cell and see chromatin, is the cell in mitosis? **No, the cell is in interphase because the chromosomes are in the form of chromatin. When the cell is in mitosis, the DNA is visible as distinct chromosomes.**

If you are in a cell during mitosis and see spindles form from centrioles, are you in a plant or animal cell and why? **You are in an animal cell because plant cells do not have centrioles.**

10.11 PROMETAPHASE

- Prometaphase is the second stage of mitosis.

Topic question

What is occurring during prometaphase? **The spindle fibers come in contact with the centromere. The chromosomes completely condense and almost line up at the exact center of the cell. The nuclear membrane completely disintegrates.**

10.12 METAPHASE

- Metaphase is the third stage of mitosis.

Topic question

What is occurring during metaphase? **The chromosomes completely line up at the equator of the cell. The spindle fibers begin to contract and tug at the sister chromatids to separate them.**

10.13 ANAPHASE

- Anaphase is the fourth stage of mitosis.

Topic questions

What happens during anaphase? **The spindle fibers contract and pull the sister chromatids apart, pulling them toward opposite ends of the cell.**

How do you know when anaphase is completed? **Each set of chromosomes are at opposite ends of the cell.**

What is karyokinesis? **The process of the chromosomes being pulled apart.**

10.14 TELOPHASE

- Telophase is the fifth and final stage of mitosis.

Topic questions

What happens during telophase? **The cell membrane pinches in the middle, which forms two daughter cells from one parent cell. The nuclear membrane reforms around the chromosomes as they start to unwind and assume the form of chromatin.**

What is the process of the cell membrane pinching inward at the middle and forming two cells called? **Cytokinesis.**

10.15 PLANT CELL DIVISION

- Plant cell division is slightly different due to the presence of the rigid cell wall.

Topic question

How does the cell plate form and what happens to it? **The cell plate is found only in plants. It forms from the Golgi or ER and eventually develops into the cell wall between daughter cells.**

10.16 ADVANCED

- The cell cycle is controlled internally by a protein called cyclin.

- When the level of cyclin reaches high enough levels in the cell, it binds to an enzyme called cyclin-dependent kinase, which activates it. Once the kinase is activated, the cell moves from interphase to mitosis.

- Normally, when cells come into contact with other cells, they stop dividing.

Topic questions

What is apoptosis? **The built-in property of cells that only enable them to only live so long. Cells routinely die as a result of apoptosis.**

Name three processes that are responsible for controlling eukaryotic cell division and maintaining homeostasis. **The cyclin level, apoptosis, and the feedback mechanism that causes cells to stop dividing when they come into contact with one another.**

10.18 KEY CHAPTER POINTS

- All cells reproduce themselves, or divide. One parent cell divides into two daughter cells.

- The idea of spontaneous generation was disproved by Redi, Spallanzani, and Pasteur.

- The Theory of Biogenesis states that living things can only come from other living things; or "life begets life".

- The purpose of cell reproduction is to create more organisms or more cells of an organism. Each newly created cell must receive a complete and fully functional copy of DNA from the parent cell.

- There are two types of cellular reproduction: asexual and sexual reproduction.

- Asexual cell reproduction results in the formation of two daughter cells that are genetically identical to one another and the parent cell.

- Mitosis, binary fission and budding are examples of cell division by asexual reproduction.

- Sexual reproduction results from the combination of genetic material from a male parent and a female parent and results in the formation of a new organism that is genetically different than either the male or female parent.

- Meiosis is an example of cell division for means of sexual reproduction.

- The cell cycle is a way of describing events of the cell's life. The cell cycle consists of interphase and mitosis. Interphase is the time when the cell is not dividing and mitosis is the time when the cell is dividing.

- The purpose of mitosis is to first create an exact copy of DNA in the parent cell and then distribute a complete copy of DNA to each daughter cell.

- DNA is replicated at the end of interphase, before the cell enters mitosis.

- DNA replication requires many enzymes, including helicase, which unzips DNA, and DNA polymerase, which copies the parent strands.

- Following DNA replication, each chromosome is attached to its identical copy at the centromere. The identical chromosome copies are called sister chromatids.

- Mitosis is broken into prophase, prometaphase, metaphase, anaphase and telophase. Each phase of mitosis is determined based upon what is happening to the chromosomes.

- Plant cell division occurs very similar to animal cell division except that plant cells do not have centrioles and a cell plate forms between the daughter cells. The cell plate later becomes the cell wall.

- Although the exact mechanisms that control the timing of the cell cycle are not well understood, it is known that the levels of cyclin, cyclin-dependent protein kinase, direct cell contact, and apoptosis all play roles in governing mitosis.

11 | Organism Reproduction: Binary Fission, Budding, and Meiosis

11.0 CHAPTER PREVIEW

In this chapter we will:

- Investigate the asexual cell/organism reproductive processes of binary fission and budding.
- Review more information regarding chromosomes and how they align during the sexual reproductive cell division process of meiosis.
- Completely define the meaning of diploid and haploid as it relates to meiosis.
- Investigate the two stages of meiosis—meiosis I and II.
- Define the differences between human male and female meiosis.
- Discuss ways in which genetic variation occurs.

11.1 OVERVIEW

- Asexual reproduction results in the production of organisms that are genetically identical, called clones.
- Organisms that are formed through sexual reproduction receive half of their chromosomes from their male parent and half from their female parent.

Topic questions

In sexually-reproducing organisms, what percent of the chromosomes come from the male parent and what percent from the female parent? **50% come from the male and 50% from the female.**

What are clones? Organisms that are genetically identical to one another.

11.2 ASEXUAL REPRODUCTION: PROKARYOTIC BINARY FISSION

- Prokaryotes reproduce through binary fission.

Topic question

Describe the process of binary fission. **Binary fission starts with DNA replication. Each copy of the DNA attaches to the cell membrane at slightly different locations. The cell begins to elongate. As it does, the two DNA molecules move away from one another. Once the cell elongates enough, cytokinesis occurs (the cell membrane pinches inward), forming two genetically identical cells.**

11.3 ASEXUAL REPRODUCTION: EUKARYOTIC BUDDING

- Many unicellular and some multicellular eukaryotic organisms reproduce asexually by budding.

Topic questions

Describe the process of budding. **A miniature version of the adult organism begins to grow from the adult. It grows by mitosis. Once it gets large enough, it falls off and is a fully-functional organism.**

Does an organism that reproduces by budding and binary fission produce organisms that are genetically different from the parent? **No, both of these processes are asexual forms of reproduction, so they form clones.**

11.4 CHROMOSOMES

- Each eukaryotic species has a certain number of chromosomes in all their cells.
- Chromosomes are arranged in pairs.

- Chromosomes are numbered to keep track of them.

Topic questions

How many chromosome number 3s do you have in your cells and where did it/they come from? **Two. All cells except sex cells have two pair of every chromosome in them. One set comes from the mom and one set from the dad.**

Why are chromosomes paired? **Each chromosome of a pair contains genes which code for the same proteins (traits). Also, one chromosome of every pair comes from the male parent and one from the female parent.**

11.5 KARYOTYPE, AUTOSOMES, AND SEX CHROMOSOMES

- A karyotype is an orderly presentation of the chromosomes of an organism.
- Autosomes are chromosomes that contain genes to determine non-sex characteristics ("body traits"). There are twenty-two pairs of autosomes in humans.
- Sex chromosomes contain genes that determine the sex of an organism. All organisms have one pair of sex chromosomes.

Topic questions

What are the sex chromosomes called? **The male chromosome is the Y chromosome and the female is the X chromosome.**

If an organism contains two X chromosomes, what sex is it? **Female.**

11.6 SEXUAL REPRODUCTION

- Sexual reproduction is the biological process in which offspring are produced by the combination of chromosomes from the male and female parents.
- A homologous pair is a matching pair of chromosomes.

Topic questions

What is a homologue? **A homologue is an individual chromosome of a pair of chromosomes.**

What is a homologous pair? **A matching pair of chromosomes.**

11.7 DIPLOID AND HAPLOID

- A diploid cell is a cell that contains a set of paired chromosomes. Somatic cells are diploid.
- A haploid cell contains a set of unpaired chromosomes. Gametes are haploid.

Topic questions

What is a diploid cell? **A cell that contains two copies of every chromosome (or a cell containing both pairs of every chromosome).**

What is a haploid cell? **A cell that contains one copy of every chromosome (or a cell containing only one chromosome of a pair).**

What happens when the chromosomes from a sperm joins with the chromosomes of an egg? **The sperm is the male gamete and is haploid (has twenty-three chromosomes). The egg is haploid (has twenty-three chromosomes). When the chromosomes are combined, a diploid organism is formed. The haploid number of the male gamete—twenty-three—and the haploid number of the female gamete—twenty-three—combine to restore the diploid number—forty-six.**

11.8 MEIOSIS

- The biological process of meiosis produces haploid gametes from diploid somatic cells.

Topic question

What is meiosis and where does it occur. **Meiosis is the biological process of making haploid gametes from diploid somatic cells. Meiosis occurs in the reproductive organs of the male and female organisms of every sexually-reproducing species.**

11.9 MEIOSIS DIVIDED

- Meiosis occurs in two steps—meiosis I and meiosis II.

Topic question

How many steps is meiosis broken into? **Two.**

11.10 INTERPHASE I

- During interphase I, the chromosomes are all copied.

Topic questions

How many total chromosomes are in a human reproductive cell after interphase I is completed? **Ninety-six.**

What is a tetrad? **A tetrad is a pair of identical homologous chromosome pairs. So a tetrad is composed of four chromosomes, two of which are identical copies as are the other two.**

11.11 PROPHASE I

- Prophase I is the first step of meiosis I.

Topic question

What are the tetrads doing during prophase I? **They are condensing and moving closer to one another.**

11.12 METAPHASE I

- Metaphase I is the second step of meiosis I.

Topic question

What is synapsis? **The process of the tetrads lining up at the equator of the cell.**

11.13 ANAPHASE I

- Anaphase I is the third step of meiosis I.

Topic questions

What occurs to the tetrads during anaphase I? **The tetrads are pulled apart from one another by the spindle.**

What happens to the diploid number during anaphase I? **When the tetrads are pulled apart, this reduces the diploid number to the haploid number.**

11.14 TELOPHASE I

- Telophase is the fourth step of meiosis I.

Topic questions

How many cells are formed after telophase I is completed? **Two.**

Are the cells formed after telophase I diploid or haploid? **Haploid.**

How many copies of chromosomes are in the cells following telophase I? **There are two copies of the haploid number in each of the two cells formed after telophase I.**

After telophase I, are the chromosomes duplicated during interphase II? **No.**

11.15 MEIOSIS II

- Meiosis II proceeds starting with prophase II, metaphase II, anaphase II, and telophase II.
- Following completion of meiosis II, there are now four cells—each of which has the haploid number (twenty-three) of chromosomes in it.
- A cell formed through meiosis is called a gamete.

Topic questions

What are the chromosomes of the pairs called in prophase II? **Since the paired chromosomes are identical copies of one another, they are properly called sister chromatids. There are twenty-three pairs of sister chromatids in the cell at this point. Even though there are twenty-three pair, or forty-six total, chromosomes in the cells during prophase II, the cells are haploid because each pair of chromosomes contains only genetic material from the father or the mother, not both. In order to be a diploid cell, one chromosome of a pair must be from the mother and one must be from the father. Students of all ages often have a hard time understanding this. You may try to explain it using two apples and two oranges. An apple and an orange represent a chromosome pair. The apple represents the chromosome from the mom and the orange the chromosomes form the dad. Together they are a diploid pair of chromosomes. When the chromosomes duplicate themselves during interphase I, there will now be two apples and two oranges. The apples stick to the apples and the oranges stick to the oranges. This means the mother's duplicated chromosomes stick to one another and the father's duplicated chromosomes stick to one another. The cell at this point is still diploid, but it has two copies of the diploid number. During telophase I, the two apples separate from the two oranges. Move the two oranges apart from the two apples so the student can see this. Now there are two cells. As the student can see, even though each cell has a pair of every chromosome in it, it is not a diploid cell because each chromosomes pair is made up of the same chromosome, not a copy of the father's and the mother's.**

What are the cells called after meiosis II is completed? **The cells are haploid gametes.**

11.16 MEIOSIS IN HUMAN MALES

- The male reproductive tissue is the teste.
- Sperm are produced in the teste.

Topic question

Following meiosis in the human male, how many sperm (gametes) are formed from one reproductive cell? **Four.**

11.17 MEIOSIS IN HUMAN FEMALES

- Meiosis is different in human females and males.
- Following completion of meiosis in females, one egg (gamete) is formed. The other three cells degenerate into polar bodies.

Topic questions

During human female meiosis, what happens to the cell that continually receives more cytoplasm during cell division? **It becomes the ovum (egg).**

How many polar bodies are formed during human female meiosis? **Three.**

11.18 ADVANCED: GENETIC VARIATION

- Genetic variation is good for a population.
- There are three common ways that genetic variation can arise—incorrect base pairing during DNA replication, crossing over, and independent tetrad alignment during meiosis I.

Topic questions

What is crossing over? **When tetrads swap a piece of chromosome material.**

When does independent tetrad alignment occur? **During metaphase I.**

11.20 KEY CHAPTER POINTS

- Budding and binary fission create new cells without combining DNA from two separate organisms. Together with mitosis, these processes represent forms of asexual reproduction. Asexual reproduction produces clones.

- Sexual reproduction creates new organisms by combining DNA from two separate organisms. Sexual reproduction requires the formation of cells specialized for this task called gametes.

- Meiosis is the specialized form of cell division that creates gametes. Gametes are the cells that fuse together and combine their DNA, creating a new organism.

- All organisms that sexually reproduce have their DNA broken into smaller segments called chromosomes. Each species has a unique number of chromosomes that all organisms of that species contain. For example, all humans have forty-six chromosomes.

- Every chromosome contains multiple gene-coding for different traits on it.

- All sexually-reproducing organisms have two sets of genes for the same trait. One set of genes is contained on one chromosome, which comes from the father; the other set is contained on a chromosome that comes from the mother. These chromosomes are called homologues

- All somatic cells are diploid (2n) while all gametes are haploid (1n or n).

- During meiosis, diploid reproductive cells generate haploid gametes through the process of meiosis.

- Meiosis is broken into two parts, meiosis I and meiosis II. Meiosis I creates cells that contain two sets of the haploid number. Meiosis II results in those cells producing gametes with one set of the haploid number of chromosomes.

- Meiosis in human males results in the production of four sperm cells (the male gamete) from one reproductive cell.

- Meiosis in human females results in the production of one ova, or egg (the female reproductive cell), and three polar bodies from one reproductive cell.

- Meiosis results in significant genetic variability between gametes from the same individual. Genetic variation is good for a species.

12 Genes and Heredity

12.0 CHAPTER PREVIEW

In this chapter we will:

- Discuss how genes are related to traits.
- Define what alleles are.
- Discuss the basic processes—such as gene segregation and independent assortment—that govern how genes are passed from generation to generation.
- Become comfortable using the technical jargon of geneticists.

12.1 OVERVIEW

- Genes are the basic unit of heredity.

Topic questions

What is heredity? **The passing of genetic information (traits) from generation to generation.**

What is the basic unit of heredity? **The gene.**

12.2 GENES

- A gene is a specific sequence on a chromosome that codes for the production of one polypeptide (protein).

Topic question

What does it mean when a gene is expressed? **When a gene is being expressed, it means the protein for which the gene codes is being made or that the trait the gene controls is visible in some way. I know we have gone over this several times, but sometimes it is hard for the student to understand.**

12.3 THE GENOME

- A genome is the entire amount of DNA of an organism.

Topic question

What is the passage of traits from generation to generation called? **Heredity.**

12.4 GENETICS

- Genetics is the study of heredity.

Topic question

What is a geneticist? **A scientist who studies genetics (or a scientist who studies heredity).**

12.5 GREGOR MENDEL

- Gregor Mendel was a monk who discovered the basic principles of genetics in the mid-1800s.

Topic questions

What did Gregor Mendel study, leading to the first understanding of heredity? **The passage of traits in pea plants.**

What did Mendel do that allowed him to meticulously follow the passage of traits? **He cross-pollinated pea plants that had the traits he wanted to study. This allowed him to control which plants passed certain traits and which plants did not pass traits.**

12.6 MENDEL DID NOT KNOW IT, BUT HE DISCOVERED ALLELES

- As a result of studying the P, F1, and F2 generations, Mendel discovered that there were two types for each of the traits he was studying.

- Mendel discovered that each trait he studied had two forms. We know now that each form is controlled by a different allele of the same gene.

Topic question

Describe what an allele is and how it is related to a gene and to the trait for which the gene codes. **An allele is an alternate form of a gene. A gene codes for a specific trait. There may be several different forms of the trait for which a gene codes. Each form of the trait is controlled by a different allele. For example, the trait of pea flower color has two forms—white and purple. There is a gene that controls the trait of flower color. The gene that codes for flower color has two alleles. One allele codes for a white flower and the other allele codes for a purple flower.**

12.7 RECESSIVE AND DOMINANT ALLELES

- A dominant allele prevents the expression of a recessive allele.

Topic question

Which trait is expressed when an organism has a dominant and recessive allele of a gene and why? **The dominant trait is expressed because the dominant trait prevents the expression of the recessive trait.**

12.8 THE GENOTYPE CONTROLS THE PHENOTYPE

- The genotype is the genetic makeup of an individual.

- The phenotype is the expression of the genotype; it is how the organism looks or functions as a result of the phenotype.

Topic questions

Let's say that a trait is coded for by two alleles. One allele, T, codes for long fingers and the other allele, t, codes for short fingers. What is the genotype and phenotype in this example? **The genotype is T and t. The phenotype is long fingers and short fingers.**

If an organism has the genotype Tt, what is the phenotype? **Since T is dominant to t (you know this because dominant alleles are always written with capital letters and recessive alleles are always written with lower case letter), the phenotype encoded by t is suppressed by the phenotype encoded by T. Therefore, an organism with the genotype Tt will have the phenotype of long fingers.**

12.9 HOMOZYGOUS AND HETEROZYGOUS CONDITIONS

- An organism that is homozygous for a condition has the same two alleles for a trait.

- An organism that is heterozygous for a condition has two different alleles for a trait.

Topic questions

What is meant when it is said an organism is homozygous dominant for a trait? **This means the organism has two dominant alleles for a condition.**

What is meant when it is said an organism is homozygous recessive for a trait? **This means that an organism has two recessive alleles for a condition.**

What does the genotype have to be for a recessive trait to be expressed? **Homozygous recessive.**

12.10 GENE SEGREGATION

- Gene segregation is the process of genes separating from one another during meiosis.

Topic questions

What happens to the alleles of the same gene during meiosis? **They separate into different gametes.**

What is this called? **Gene segregation.**

What will the genotype of the gametes be for an organism whose phenotype if Bb? **One of the gametes will be B and the other will be b.**

12.11 GENE SEGREGATION AND THE F2 GENERATION

- Gene segregation explains the appearance of the recessive trait in the F2 generation.

Topic question

When a heterozygous organism undergoes meiosis, what is the genotype of the resulting gametes? **One gamete will have the dominant gene and one will have the recessive gene.**

12.12 PUNNETT SQUARES

- A Punnett square is a tool used to follow genotypes and phenotypes from generation to generation. It relies on the property of gene segregation.

Topic question

Can a Punnett square be used to predict expected genotype and phenotypes of offspring? **Yes. One of the major uses of Punnett squares is the ability to predict expected genotypes and phenotypes of offspring.**

12.13 INDEPENDENT GENE ASSORTMENT

- During gene segregation, genes segregate independent of one another.
- Mendelian principles are the principles of gene assortment that were first discovered by Mendel.

Topic question

What does independent assortment mean? **When genes segregate during meiosis, they separate independently. This means they are distributed to gametes without relationship to one another.**

12.15 KEY CHAPTER POINTS

- Genes are the smallest unit of heredity. They are passed from parent to offspring through sexual or asexual reproduction.
- When a gene is expressed, it means that the gene is being transcribed into mRNA, and the mRNA is being translated into the gene's protein. Basically, it means that whatever trait the gene codes for will be active or visible.
- The entire amount of DNA in any given organism is referred to as the genome.
- Geneticists study the passage of genes from parent to offspring and learn to predict the patterns of the traits being passed. The passage of genes (traits) from parent to offspring is called heredity; therefore, geneticists study hereditary patterns.
- Gregor Mendel laid the foundations of modern genetics by studying the passage of traits in pea plants.
- Alleles are forms of the same gene that code for a slightly different trait.
- The simplest relationship genes can have with one another is dominant/recessive, which is the relationship Mendel was studying.
- Genotype is the genetic makeup of an individual; phenotype is the way the genotype is expressed. Another way of stating this is the genotype are the genes and the phenotype is the trait the genotype creates.
- Gene segregation splits up alleles so that each gamete receives one allele for each gene.
- Punnett squares are commonly used in genetics to follow and predict hereditary patterns.

13 | Inheritance Patterns

13.0 CHAPTER PREVIEW

In this chapter we will:

- Discuss the importance that probability plays in the study of genetics.

- Investigate different inheritance patterns and use examples to understand how they relate to genetic diseases.

13.1 OVERVIEW

- Inheritance patterns refer to the ways in which traits (genes) are inherited from generation to generation.

Topic question

What is one of the most important things to realize about inheritance patterns? **How little we actually know about the ways in which traits are passed from generation to generation.**

13.2 PROBABILITY

- Probability is the measure of how likely an event is to occur.

Topic question

If you are holding a six-sided dice, what is the probability that you will roll a six? **The probability of that is equal to the outcome desired (rolling a 6) divided by the total possible outcomes (rolling 1, 2, 3, 4, 5 or 6. That means there are six possible outcomes. The likelihood of rolling a 6 is 1/6. Other ways of stating that is a one-in-six chance or a 16.7% chance.**

13.3 PROBABILITY AND INHERITANCE

- We can apply principles of probability to gene segregation because gene segregation occurs randomly.

- When looking at the probability of successive events occurring, one must multiply the probabilities of the individual events together to obtain the probability of the successive events occurring.

Topic question

What is the probability of rolling three successive 6s on a six-sided dice? **The probability of successive events occurring is equal to the product of the three independent events occurring. Since the probability of getting one 6 is 1/6, the probability of rolling three successive 6s is 1/6 x 1/6 x 1/6, or 1/216. Therefore the chance of this happening is 1/216, or 1 in 216 or 0.5%.**

13.4 INHERITANCE PATTERNS: INCOMPLETE DOMINANCE

- Incomplete dominance is displayed when there is no one allele that is dominant over the other.

Topic question

There are two alleles, one coding for black fur and one coding for white fur. These alleles exhibit an incomplete dominance type of inheritance pattern. What does that mean and what would an animal look like if it had one black allele and one white allele for fur color? **Incomplete dominance means there is no form of an allele for a certain gene which is dominant** over the other and the resulting phenotype is a blend of the phenotype of each gene. In this case, with the animal having one allele for black fur and one allele for white fur, the fur would be a blend of black and white, or appear gray.

13.5 INHERITANCE PATTERNS: MULTIPLE ALLELES AND CO-DOMINANCE

- Some genes have more than two alleles. These are called multiple alleles.
- Some alleles do not show a dominant/recessive or an incomplete dominance pattern. Instead, the traits of both alleles are expressed. This is called co-dominance.

Topic question

Let's say the trait for seed color in sunflowers is controlled by one gene. There are four different alleles that code for this trait, each of which results in different seed colors. One of the four alleles—we will call it L—codes for black seeds. Another of the four alleles—R—codes for seeds that are black with white spots. Plants that are LL make black seeds. Plants that are RR make seeds that are black with white spots. Plants that are LR make both seeds that are black and seeds that are black with white spots. What type of inheritance pattern is this? **Co-dominance.**

13.6 INHERITANCE PATTERNS: SEX-LINKED TRAITS AND DISEASES

- Sex-linked traits are traits that are not related to the sex of an organism, but are carried on the sex chromosomes.
- The presence of sex-linked traits was first discovered in the fruit fly in the early 1900s.

Topic questions

Which chromosome carries most of the sex-linked traits? **The X chromosome.**

Which sex chromosome does a boy inherit from his mother? **The X chromosome. The only way for a boy to be a boy is to inherit the Y chromosome from his dad. Therefore, by default the X chromosome comes from his mom.**

Is it impossible for a girl to have a recessive sex-linked trait? **No. It is unusual, but not impossible. As shown by the fruit fly experiments, a female can have a recessive sex-linked trait if she inherits a recessive sex-linked gene from her mom and the same recessive sex-linked gene from her dad.**

13.7 INHERITANCE PATTERNS: CONTINUOUS VARIATION

- Continuous variation is an inheritance pattern in which the phenotype is controlled by modifier genes. In continuous variation, the phenotypes fall within a range of extremes.

Topic question

Is human hair color an example of continuous variation? Why or why not? **Yes, it is. There are all different shades of hair color that fall within standard accepted ranges. Non-colored hair is an excellent example of continuous variation.**

13.8 INHERITANCE PATTERNS: EPISTASIS

- Epistasis is the interaction between a gene that codes for a trait and one or more genes that do not code for a trait so that the expression of the gene which does code for the trait is modified (often suppressed).

Topic question

Explain epistasis using the mouse-coat color example. **Mice have two coat colors, black and brown. Black is dominant to brown. A mouse that is homozygous for the black-coat color, or heterozygous, will have a black coat. A mouse that is homozygous for the brown color will have a brown coat. However, there is another gene that determines whether the pigment will deposit in the hair or not. The allele for deposition of pigment in the hair is dominant to the allele for no deposition of pigment in the hair. A mouse that is homozygous for the recessive non-deposition allele will appear white no matter what the genotype is for coat color. The gene for deposition of coat pigment is a modifier, or is said to be epistatic, to the gene which codes for coat color.**

13.9 INHERITANCE PATTERNS: ENVIRONMENT

- The phenotype can also be affected by the environment.

Topic question

Explain environmental factors and the influence they can have on skin color. **If a person with the genotype for light-colored skin moves to the tropics and is in the sun all the time, their skin will tan and turn much darker than their genotype codes for. The reason their skin becomes darker is not because their genes have changed. It is because of the environmental change they are exposed to. The more sun you are exposed to, the darker your skin gets (i.e. it tans). The opposite is true for a person who has lived in the tropics all their life who moves to a northern climate. Since they do not get as much sun, their darker skin becomes lighter and appears the skin color for which they are genetically programmed.**

13.11 KEY CHAPTER POINTS

- The relationship of genotype and phenotype is usually more complex than dominant-recessive relationships.

- Probability is predicting how likely an event is to occur. Probability is commonly used in genetics to predict how likely it is that offspring may inherit a certain trait or disease from their parents.

- Incomplete dominance is an inheritance pattern in which no allele for a particular gene is dominant or recessive. Traits controlled by alleles showing incomplete dominance demonstrate a blend of the phenotypes of the two traits.

- Some traits are controlled by genes with more than two alleles. This situation is called multiple alleles.

- Codominance is an inheritance pattern in which neither allele for a gene is dominant over the other. Codominant traits result in the organism expressing the phenotypes of both genes equally.

- Sex-linked traits are non-sex, related traits that are carried on the sex chromosomes.

- Continuous variation is an inheritance pattern in which the phenotypes fall within a range rather than in exact categories.

- Epistasis is the influence of one or more genes on phenotype expression for a trait (or traits) coded for by another gene.

- Environmental conditions can alter the phenotype.

14 | Genetic Variation

14.0 CHAPTER PREVIEW

In this chapter we will:

- Discuss normal genetic variation and how it is generated.
- Investigate point mutation and chromosomal mutation types.
- Learn the difference between somatic and germ cell mutations.

14.1 OVERVIEW

- There are normal and abnormal types of genetic variation.

Topic question

How do different alleles for the same gene arise? **Through mutations that alter the sequence of the gene in some way.**

14.2 NORMAL GENETIC VARIANCE

- Normal genetic variation is critical for the survival of a population.
- Normal genetic variation occurs as a result of alleles.
- Normal genetic variation occurs by independent assortment and crossing over.

Topic questions

When is normal genetic variation usually produced? **During meiosis.**

What are the ways in which normal genetic variation can occur? **Independent assortment and crossing over.**

14.3 ABNORMAL GENETIC VARIANTS: MUTATION

- Mutations are a change in the normal sequence of a gene.
- Almost all mutations cause harm or death to the organism.

Topic question

What is a change in the normal sequence of DNA called? **Mutation.**

14.4 DNA: THE MOLECULE REVISITED

- Genes make up no more than 1% of the nucleotide sequence of a chromosome. The rest is taken up by RNA genes, tandem repeats, pseudogenes, and introns.

Topic question

Most of the space on DNA is taken up by genes. True or False and why? **False. Almost all the space of DNA is taken up by non-gene sequences. No more than 1% of DNA actually contains nucleotide sequences that code for the production of a protein.**

14.5 MUTATION

- Mutations arise from a number of different mechanisms.
- Almost every mutation that has been observed is bad for the organism.
- Germ cell mutations are passed from generation to generation.
- Somatic mutations are not passed from generation to generation.

• Mutations can affect one nucleotide only, or can affect larger segments of chromosomes.

Topic questions

What is a germ cell mutation? **A mutation that occurs and is passed into a gamete. Germ cell tumors can be passed from parent to offspring.**

Remembering our discussion of probability from the last chapter, if a mutation occurs, is it more likely to occur in a gene or in a non-gene segment of DNA? **In a non-gene segment. Since the portion of DNA that does not code for a gene is so much larger than the portion that does, it is much more likely that a mutation will occur in a non-gene segment.**

14.6 POINT MUTATION TYPES

• A point mutation results in the replacement of a single nucleotide with another one.

• Point mutations are broken into substitutions, deletions, and additions.

• Silent mutations are mutations that occur but are never noticed

Topic questions

What are the types of point mutations? **Substitutions, deletions, and additions.**

What type of mutation has occurred when a nucleotide is left out of its normal position? **Deletion mutation.**

Is it possible for a substitution mutation to occur and not be recognized? If not, why not? If so, how? What is it called? **A substitution mutation can occur that does not cause any observable change in the protein or the function of the protein that the gene codes for. These are called silent mutations. For example, both AAA and AAG code for the amino acid phenylalanine. If the third A is the AAA codon undergoes a substitution mutation so that the A is changed to a G, then the codon still codes for phenylalanine. Another situation can occur in which a substitution mutation can occur that alters a codon in such a way that the amino acid is changed. However, if the new amino acid has similar properties to the original amino acid, then the protein may still function normally.**

14.7 MISSENSE MUTATIONS

• Missense mutations are caused by substitutions.

• A missense mutation results in an improper amino acid being placed into the protein during protein synthesis.

Topic question

What is a missense mutation and how does it occur? **A missense mutation occurs as a result of a substitution-point mutation. A missense mutation changes the codon and results in the insertion of the incorrect amino acid into the protein.**

14.8 NONSENSE MUTATIONS

• Nonsense mutations are caused by substitution-point mutations.

• A nonsense mutation alters an amino acid encoding codon into a stop codon.

Topic question

What is a nonsense mutation? **It is a substitution mutation that changes an amino acid encoding codon to change into a stop codon.**

14.9 POINT MUTATIONS, THE READING FRAME, AND FRAME SHIFT MUTATIONS

• Substitution mutations do not alter the reading frame.

• Deletion mutations and addition mutations do alter the reading frame.

• Frame shift mutations result from deletion or addition point mutations.

• Frame shift mutations are usually much more harmful to the organism than substitution mutations.

Topic questions

Why do deletions and additions alter the reading frame? **The reading frame is maintained by the orderly linear placement of nucleotides. When an addition or deletion mutation occurs, a nucleotide is either added or removed from its linear position. As a result, the reading frame changes. This is a difficult concept to put into words, but if the student can do it, they really understand this material well! They could also use an example to show how the reading frame changes. I have several examples in the book they can use.**

What type of mutation has occurred to this sentence? Normal sentence: THE DOG RAN AND ATE THE PIG   Mutated sentence: THE DGR ANA NDA TET HEP IG **This is an example of a deletion mutation. The "O" in "DOG" was deleted.**

Why is the sentence so hard to read following the mutation? **The reading frame has shifted and it does not make any sense.**

14.10 CHROMOSOMAL MUTATION

- Chromosomal mutations involve much larger portions of DNA than point mutations.
- Most chromosomal mutations occur during mitosis and meiosis.

Topic question

What type of mutation involves larger portions of DNA, point mutations or chromosomal mutations? **Chromosomal.**

14.11 CROSSING OVER

- Crossing over occurs when homologous chromosomes swap pieces of their DNA during meiosis.

Topic question

Usually crossing over is not harmful to the organism because the swapping of DNA is fairly even between the two chromosomes. True or False? **True. Usually similar or the exact gene segments are involved in the crossing over.**

14.12 CHROMOSOMAL DELETION MUTATION

- A chromosomal deletion mutation occurs when an entire segment of a chromosome is lost.

Topic question

A chromosomal deletion is not a big deal for the organism and usually does not cause any harm. True or False? **False. Chromosomal deletions usually result in the death of the organism.**

14.13 CHROMOSOMAL INVERSION MUTATION

- An inversion mutation involves a piece of a chromosome breaking off and then reinserting back into the same chromosome, but in the wrong place.

Topic question

Why may an inversion mutation not harm an organism? **Depending on where the chromosome piece breaks off and reinserts, gene segments may not be involved. As long as genes are not disrupted as a result of the inversion, there will likely be no harmful effect to the organism.**

14.14 CHROMOSOMAL DUPLICATION MUTATION

- A duplication mutation occurs when a piece of one homologous chromosome breaks off and is inserted into the other homologous chromosome.

Topic question

When a duplication mutation occurs to one chromosome, what has to happen to the homolog that donated the material for the duplication? **Since one chromosome gains a segment of another chromosome, the chromosome that donated the segment has undergone a deletion mutation. I want the student to realize this material has to come from somewhere!**

14.15 CHROMOSOMAL TRANSLOCATION MUTATION

- A translocation mutation involves a segment of chromosome breaking off of one chromosome and inserting into a non-homologous chromosome.

Topic question

What is a translocation mutation? **A translocation mutation involves a segment of chromosome breaking off of one chromosome and inserting into a non-homologous chromosome.**

14.16 NONDISJUNCTION CHROMOSOMAL MUTATION

- Nondisjunction mutations occur when homologous chromosomes do not separate during karyokinesis.

Topic question

What is the basic problem of a nondisjunction mutation? **The chromosomes do not separate properly during karyokinesis; they "stick" together. This results in one gamete receiving two chromosomes of a pair and the other gamete receiving none.**

14.17 MUTAGENS

- Mutagens are environmental factors that are known to cause mutations.

Topic question

List three known mutagens. **Cigarette smoke, cigarette smoke, and cigarette smoke. Also x rays, UV rays, and certain chemicals are all known to cause changes in DNA which lead to cancer.**

14.18 ADVANCED

- Germ cell mutations are passed from parent to offspring. Somatic cell mutations are not.

Topic question

Why are germ cell mutations passed from parent to child? **Germ cell tumors occur in the gamete. This means the gamete is formed with a mutation present. All cells of an individual grow from the single cell formed when a male gamete fuses with a female gamete. If one of those gametes has a mutation, then the cell formed when that gamete fuses with another gamete will have the mutation. As the single cell undergoes mitosis to grow in size from a single cell to a multicellular organism, that mutation is passed to all the cells every time a cell divides. Then when the organism is an adult, assuming it lives that long, half the gametes that the organism produces will also have the mutation in them, so the mutation can be passed indefinitely.**

14.20 KEY CHAPTER POINTS

- Sometimes mutations arise in the DNA and genes during meiosis or mitosis. Mutations are changes in the normal sequence of DNA.
- Normal genotype variances arise through the process of independent assortment and crossing over during meiosis.
- Abnormal genetic variances occur due to mutation.
- Mutations are almost always fatal or, at best, detrimental to the organism that inherits the mutated gene(s).
- It is estimated that 1% of an organism's DNA sequences are actual genes. The remainder of the DNA codes RNA, ribosomes, tandem repeats, and pseudogenes.
- Point mutations result in a change of only one nucleotide in DNA. There are three types of point mutations: addition, deletion, and substitution.
- Addition and deletion mutations cause frame-shift mutations.
- Missense mutations are point mutations that result in the insertion of the wrong amino acid into the protein for which the gene codes.

- Nonsense mutations are point mutations which result in the insertion of a stop codon into the gene. This leads to premature termination of protein translation.

- Chromosomal mutations occur when there are large segments of chromosomal material gained or lost by a chromosome. The types of chromosomal mutations are: crossing over, deletion, inversion, duplication, translocation, and nondisjunction.

- Mutagens are things that are known to cause genetic mutations.

15 | Human Genetics

15.0 CHAPTER PREVIEW

In this chapter we will:

- Define congenital genetic diseases.

- Learn why most genetic diseases are inherited as an autosomal recessive pattern.

- Discuss the meanings of carriers, affected, and unaffected individuals as they relate to genetic disorders.

- Learn how to use a pedigree analysis.

- Investigate:
    - the autosomal recessive genetic diseases of phenylketonuria, cystic fibrosis, and sickle cell anemia.
    - the autosomal dominant genetic disease of Huntington's chorea.
    - the X-linked diseases of hemophilia and color blindness.
    - the polyploid condition of Down's syndrome.

15.1 OVERVIEW TO CONGENITAL DISEASES

- Congenital genetic diseases are those abnormalities that people are born with and are due to abnormal genes.

- Many congenital diseases are called syndromes because the chromosome abnormalities result in common problems associated with the specific chromosomal mutation.

Topic question

What is a congenital genetic disease? **A disease a child is born with due to a mutation in the DNA.**

15.2 UNAFFECTED, CARRIERS, AND AFFECTED

- Affected persons have a genetic disease or condition.

- Carriers do not have a genetic disease or condition, but carry the gene which causes the disease or condition.

- Unaffected persons do not have a disease or condition and do not carry the gene for it either.

- Most genetic diseases are transmitted in an autosomal, recessive fashion.

- Some genetic diseases are transmitted in a sex-linked fashion.

Topic questions

What is a carrier of a recessive condition? **A carrier is someone who has one normal/ dominant allele and one recessive allele for a condition or disease.**

Does a carrier exhibit the recessive condition they carry the allele for? **No. Since a carrier has one normal/dominant allele, its trait is expressed, so there is no effect of the recessive allele on a carrier.**

If neither a man nor his wife are carriers for a disease, could they pass along the allele for the disease to their offspring? **Yes. This is kind of a trick question. If neither parent is a carrier, then both parents are either affected by a disease, both parents are unaffected by a disease, or one parent is affected and one is unaffected. If one or both parents are affected by a disease, then they will definitely pass along the recessive allele to their offspring.**

What does it mean when a genetic disease is said to be transmitted in an autosomal recessive fashion? **It means that the allele for the disease is carried on one of the somatic chromosomes, or autosomes, and that the allele is a recessive allele, so the person needs to inherit two recessive alleles to have the disease.**

15.3 PKU

- PKU is a genetic disease transmitted in an autosomal recessive fashion.

Topic questions

Is PKU the wild type phenotype? **No, it is the non-wild type (or mutated) phenotype.**

How is PKU inherited? **In an autosomal, recessive fashion.**

15.4 PEDIGREE ANALYSIS

- A pedigree analysis is a tool used to track inheritance of traits through many generations.
- Pedigree analyses are particularly useful to track autosomal recessive and sex-linked diseases in families.

Topic question

What is a pedigree analysis? **A pedigree analysis is a tool used to follow the passage of traits from one generation to the next. It shows the phenotypes of parents and their offspring from generation to generation. They are particularly useful in following the passage of recessive traits and diseases in families.**

15.5 CYSTIC FIBROSIS

- Cystic fibrosis is an autosomal recessive genetic disease resulting in increased airway and GI tract secretions, recurrent infections, and breathing problems.
- Cystic fibrosis is caused by a defective membrane channel protein.

Topic questions

What is the inheritance pattern of cystic fibrosis? **Autosomal recessive.**

What is the particular mutation and what is the result of the mutation? **The mutation is in the gene that codes for a chloride channel membrane protein. This results in a protein that does not move chloride ions in and out of lung and pancreas cells properly.**

15.6 SICKLE CELL DISEASE

- Sickle cell disease is inherited as an autosomal recessive disease and results in an inability to carry oxygen to the tissues properly.
- Sickle cell disease is caused by a mutation in one of the proteins that makes up hemoglobin.

Topic questions

What is the mutation in sickle cell disease? **There is a substitution mutation in the sixth codon of one of the protein chains of hemoglobin. This results in the amino acid valine being inserted into this chain instead of the normal glutamic acid.**

Why does the hemoglobin molecule not work properly in sickle cell disease? **The correct answer is not because there is a mutation in one of the genes. The correct answer is because the replacement of glutamic acid with valine results in a hemoglobin molecule with a abnormal structure distorting the shape of red blood cells. As a result the hemoglobin cannot carry oxygen properly.**

15.7 HETEROZYGOUS SUPERIORITY

- The condition of the heterozygous form being protective against some disease or illness is called **heterozygous superiority**.

Topic question

What is heterozygous superiority? **It is the protection that the heterozygous condition can give to people who are carriers of a recessive allele.**

15.8 HUNTINGTON'S DISEASE

- Huntington's disease is an autosomal dominant genetic disease.

Topic questions

What is an autosomal dominant genetic disease? **It is a disease caused by a mutation on a somatic chromosome that is transmitted in a dominant fashion. Unlike most genetic diseases, it is caused by a dominant gene.**

If a person who has Huntington's disease has a child, what is the probability the child will inherit the gene for Huntington's? **The probability is 50%. When the parent with Huntington's makes gametes, there is a 50% chance that a gamete will receive the allele for Huntington's. Since the disease is transmitted in a dominant fashion, the child has a 1 in 2 chance of receiving the gamete with the dominant Huntington's allele.**

15.9 SEX-LINKED DISEASES

- Most sex-linked disease are transmitted on the X chromosome; therefore, most sex-linked diseases are called x-linked diseases.
- Hemophilia and color blindness are x-linked diseases.

Topic question

Why are most people affected by x-linked diseases male? **Almost all sex-linked diseases are transmitted on the X chromosome. Since males only have one X chromosome, they do not have the possibility of having a normal gene on their other sex chromosome. The Y chromosome does not carry many non-sex determining genes. Therefore, if a boy inherits an X chromosome from his mom that has a recessive gene on it for a disease, he will have the disease, because there is not a normal, dominant gene to counteract it.**

15.10 ANEUPLOIDY

- Aneuploidy means having the wrong number of chromosomes in the cells.
- Polyploidy means having too many chromosomes in the cells.
- Aneuploidy is caused by nondisjunction during meiosis.

Topic questions

What is aneuploidy caused by? **It is caused by nondisjunction during meiosis.**

Is trisomy 21 a polyploid condition? **Yes. A polyploid condition is one in which there are too many chromosomes in the cells. Trisomy 21 cells have three, instead of two, chromosome number 21 in them.**

15.11 ADVANCED

- Chorionic villus sampling, fetoscopy, and ultrasound are three ways that a fetus can be assessed for the presence of a genetic disease.

Topic question

What is fetoscopy? **It is a medical procedure in which a fetus is viewed with a small camera.**

Christian perspective

Is there any need for fetal imaging in a Christian viewpoint?

15.13 KEY CHAPTER POINTS

- Congenital genetic diseases are diseases which are transmitted from parent to offspring. Most congenital diseases are cause by recessive genes.

- Affected individuals are said to have a genetic disease or recessive trait. A disease or trait caused by recessive alleles requires that the affected individual have both recessive alleles in order to have the condition.

- Carriers of a recessive condition have one recessive allele and one dominant allele for the condition for which the gene codes.

- Unaffected individuals have two dominant alleles for the condition for which the gene codes.

- A pedigree analysis is a way of tracking traits through multiple generations in a family.

- Cystic fibrosis is the most common inherited genetic disease among Caucasians. It is transmitted as a recessive condition.

- Sickle cell disease is the most common inherited genetic disease among people of African descent. It is transmitted as a recessive condition.

- Recessive alleles exist due to heterozygous superiority.

- Huntington's disease is transmitted in a dominant fashion.

- Sex-linked diseases, such as hemophilia and color-blindness, are carried on the X chromosome. They also follow a recessive hereditary pattern, but since males only have one X chromosome, they will have the disease if they inherit the recessive allele from their mother.

- Aneuploidy is the condition of having the incorrect number of chromosomes. The most common polyploid condition is trisomy 21.

- There are a number of ways to test a fetus for the presence of genetic diseases. These include fetal ultrasound, amniocentesis, and fetoscopy. However, there are serious ethical and moral questions to these practices regarding what is done with the results.

16 | DNA Technology

16.0 CHAPTER PREVIEW

In this chapter we will:

- Define the terms associated with DNA technology.
- Discuss the process of transformation.
- Investigate the common methods for making recombinant DNA and inserting it into appropriate hosts.
- Review several ways in which recombinant DNA technology is being applied on a daily basis.

16.1 OVERVIEW

- Genetic engineering is the technique of applying all our knowledge regarding DNA technology for practical purposes.

Topic question

Why is genetic engineering considered an area of great hope for treating diseases, especially congenital genetic diseases? **Many diseases may be able to be cured by giving the normal gene to a person with a disease caused by a defective gene.**

16.2 TRANSFORMATION

- Transformation is the genetic alteration of a cell resulting from the uptake and expression of foreign DNA.

Topic question

Describe the process of transformation which Dr. Griffith performed with *Streptococcus* bacteria. **There are two forms of the bacteria *Streptococcus pneumoniae (Strep pneumo). Only the smooth form of *Strep pneumo* cause disease. The rough *Strep pneumo* do not cause sickness. Dr. Griffith injected the smooth *strep* bacteria into mice and they all got sick and died. He isolated the smooth *strep* from their blood. He then injected the rough *strep* into mice and they did not get sick or die. He then took the smooth *strep* and heated them to the point that they all were dead. He took these dead smooth *strep* and mixed them with live rough *strep* and injected the mixture into the mice. Many of the mice became sick and died and he isolated smooth *strep* from the blood of those that died. From this experiment, he deduced that some part of the dead smooth bacterial cells, or some molecules from the dead cells, were able to make the non-disease-causing rough bacteria change into disease-causing smooth bacteria. To identify what it was, the smooth cells were completely ground up and allowed to settle. The fluid was separated from the more solid cell components. Only molecules were contained in the fluid as the cell parts were left out. The fluid was then added to a live culture of rough cells. The rough cells were able to reproduce in this environment for several generations, and soon smooth bacterial colonies were growing. This proved that some type of chemical or molecule was the factor that caused the rough cells to develop the smooth trait. These smooth cells were then injected into the mice and the mice became sick and died. We now know the reason the cells transformed from non-disease-causing cells to disease-causing cells is because they took up the DNA from the smooth cells. The DNA contained the gene that coded for the smooth-cell trait. When this DNA was taken up by the rough cells, they started to make the smooth protein. Once they started to make the smooth protein, they were able to infect the mice.**

16.3 RECOMBINANT DNA

- Recombinant DNA is the name given to one molecule of DNA which contains DNA from two or more organisms.

Topic question

A scientist takes DNA from a firefly and inserts it into a tobacco plant. The piece of DNA he took from the firefly contains the gene that codes for an enzyme called luciferase. Luciferase catalyzes the reaction and makes fireflies glow. Now the tobacco plants glow as a result of this (this has been done with a number of organisms). What is the DNA called that the tobacco plant contains? **It is recombinant DNA because the DNA in the plant is composed of DNA from two organisms—the firefly and the tobacco plant.**

16.4 REVERSE TRANSCRIPTASE

- Reverse transcriptase is an enzyme that can make a single-stranded DNA molecule using an mRNA template.

Topic questions

Describe how reverse transcriptase is used to make a molecule of double-stranded DNA. **Reverse transcriptase is added to the mRNA that codes for insulin, along with the nucleotides A, T, C, and G. After a while, a single-stranded DNA was made. This single-stranded DNA molecule represents the gene for the insulin. The single-stranded DNA molecule is then mixed with DNA polymerase and more A, T, G, and C. After a while, the single-stranded DNA would be made by the DNA polymerase into a double-stranded DNA molecule.**

Why is using the reverse transcriptase method called backwards recombinant technology? **Because the gene is obtained from making a DNA molecule from an mRNA template. Usually mRNA is made from a DNA template, so this method is called backwards.**

16.5 BRIEF OVERVIEW OF MODERN RECOMBINANT TECHNOLOGY

- Current methods of making recombinant DNA use enzymes to cut DNA into smaller pieces, a vector and an appropriate host.
- Most current methods of making recombinant DNA use the direct gene that codes for the protein or enzyme of interest rather than the mRNA/reverse transcriptase method.

Topic question

What are the three basic things needed to make recombinant DNA? **Restriction enzymes, a vector, and a host.**

16.6 RESTRICTION ENZYMES

- Restriction enzymes cut DNA at specific nucleotide sequences.
- Restriction enzymes are used to cut DNA and make sticky ends.

Topic question

What are two important features of restriction enzymes? **They cut DNA at specific nucleotide sequences; when they cut the DNA, sticky ends are made.**

16.7 VECTORS

- A **vector** is a unit that introduces DNA into a host.
- The two most common vectors used are viruses and plasmids.

Topic questions

What are the two most common vectors used in DNA technology? **Viruses and plasmids.**

How is a recombinant vector made? **The DNA containing the gene is cut by the desired restriction enzyme. This results in the linear DNA molecule with sticky ends. The vector DNA is cut with the same restriction enzyme. This also makes a linear DNA molecule with sticky ends. The sticky ends of the gene are complimentary to the sticky ends of the vector. An enzyme called ligase is then added to the cut gene DNA and the cut vector DNA. Ligase is an enzyme that links together DNA nucleotides that are complimentary to one another. When the vector and gene DNA are mixed together with the ligase, ligase links together the sticky ends of the gene DNA to the sticky ends of the vector DNA. This forms one circular piece of recombinant DNA. Now a** recombinant plasmid has been made. This process is repeated over and over until thousands or millions of plasmids have been made, each containing the same gene.

16.8 GETTING THE RECOMBINANT DNA INTO THE HOST

- A host is a suitable organism to receive the recombinant DNA.
- Virtually any organism can be a host.

Topic questions

What is a host? **A suitable organism into which recombinant DNA can be inserted.**

What is a transgenic organism? **An organism that has had DNA introduced into it from another species using a vector.**

16.9 RFLP ANALYSIS AND DNA FINGERPRINTING

- RFLP analysis is a way to cut the huge DNA molecules into smaller pieces to analyze them.
- DNA fingerprints are the bands that are made by restriction fragments when they are electrophoresed.

Topic question

Describe the process of DNA fingerprinting. **Large segments of DNA are first cut by restriction enzymes. This creates restriction fragments. The restriction fragments are placed onto a gel and electrophoresed. This separates the fragments based on their size and electrical charge. It creates bands of DNA in the gel. A special camera takes a picture of the gel after some radioactive material has been added, which sticks to the DNA fragments.**

16.10 POLYMERASE CHAIN REACTION

- PCR is a technology used to make a lot of DNA from a sample that contains little DNA.

Topic question

What is the DNA technology used to make a lot of, or amplify, a small amount of DNA? **Polymerase chain reaction.**

16.11 APPLIED DNA TECHNOLOGY: GENE ISOLATION

- The earliest application of gene technology has been in isolating a gene.

Topic question

16.12 APPLIED DNA TECHNOLOGY: SELECTIVE BREEDING

- Selective breeding is a process by which organisms containing a particular trait, or traits, are bred with one another across many generations in order to make that trait more obvious.

Topic question

Describe the process of selective breeding if you were a farmer who wanted to breed fish to have large eyes. **The fish in question with the largest eyes would be collected and bred with one another. Over the generations, only the fish with the largest eyes would be allowed to breed with one another. Over time, the eyes would gradually become larger and the fish that had the largest eyes were bred with one another.**

16.13 APPLIED DNA TECHNOLOGY: GENE CLONING

- Recombinant technology is used to manufacture insulin for medical use.
- Recombinant technology is also used to make vaccines.

Topic question

Recombinant technology is of no practical use in today's world. True or False? **False. Recombinant technology is commonly used to make vaccines and medications.**

16.14 APPLIED DNA TECHNOLOGY: GENE THERAPY

- In gene therapy, a person born with a defective gene is given the proper gene through recombinant technology, hopefully curing the genetic disease.

Topic question

Describe how one could possibly cure cystic fibrosis with gene therapy. **Since cystic fibrosis is a disease caused by a defective gene, being able to replace that defective gene with a normal gene could possibly cure this disease. When the normal gene is given to the pancreas and lung cells, it would allow these cells to start making the normal membrane channel protein, which pumps chloride.**

16.15 APPLIED DNA TECHNOLOGY: FORENSICS

- DNA technology is commonly used to investigate crime scenes, especially if there is scant physical evidence.

Topic question

Which DNA technology is particularly useful in forensics? **DNA fingerprinting.**

16.16 ETHICS

Topic question only

Do you think "cloners" are playing God?

16.17 KEY CHAPTER POINTS

- Genetic engineering is the application of molecular genetics and DNA technology for practical purposes.
- Transformation is the genetic alteration of a cell by introduction, uptake, and expression of foreign DNA into a cell. This process has the ability to change or introduce new traits into a cell.
- Recombinant DNA is DNA made up of genetic material from two or more unrelated organisms.
- Reverse transcriptase is an important enzyme in recombinant DNA technology. It makes DNA from an mRNA template.
- A vector is used to introduce recombinant DNA into an appropriate host.
- Restriction enzymes are used to cut DNA at precise locations.
- Common vectors are plasmids and viruses.
- Common host organisms are bacteria, particularly *E. coli*.
- Recombinant DNA can be used to make large amounts of a particular gene product (protein). It can also be used to introduce normal genes into organisms which contain abnormal genes.
- Gene therapy may be able to cure genetic diseases such as cystic fibrosis and muscular dystrophy. Gene therapy also has the potential to cure cancer and diabetes.
- DNA technology of RFLP analysis, which makes a DNA fingerprint, can be used to identify people and to link criminals to crime scenes.
- There are serious moral and ethical issues which need to be addressed regarding the use of DNA technology.

The Origin of Life
17 and the Fossil Record

17.0 CHAPTER PREVIEW

In this chapter we will:

- Discuss the two major philosophies for the origin of life on earth—evolution and creation.
- Establish that evolutionary thought is founded on the basis of atheism.
- Establish that creation is founded on the basis of an all-powerful, all-knowing God who created the universe and everything in it.
- Investigate how a person's philosophy dictates the way they interpret scientific data.
- Examine the fossil record and the evolutionary and creationist interpretation thereof.
- Investigate evolutionary and creationist viewpoints regarding strata formation.
- Discuss the practices and pitfalls of radiodating techniques.

17.1 OVERVIEW

- Evolution is the scientific philosophy that all life on earth is here by accident; that over many billions of years, less complex organisms have acquired new traits, by chance and changed into new organisms. Evolution is the belief system that all life on earth came from a single-celled bacterium.

Topic question

What is evolution? **Evolution is the scientific philosophy that all life on earth is here by accident. That over many billions of years, less complex organisms have acquired new traits, by chance, and changed into new organisms. Evolution is the belief system that all life on earth came from a single-celled bacterium.**

17.2 EVOLUTION: BASIC PHILOSOPHY

- Evolutionists believe in evolution.
- According to evolutionists, the earth is about five billion years old.
- Evolutionists believe that life arose on the planet by pure chance alone.

Topic question

What is the basic process evolutionists believe is responsible for the origin of life? **Evolutionists believe that life was able to form by pure chance by lightning electrifying bodies of water that contained pre-organic molecules. The pre-organic molecules organized into organic molecules, then the organic molecules organized into proteins, lipids, carbohydrates, and DNA. Eventually a single cell organism was formed. Over billions of years, all life forms on earth came from that single-celled organism.**

17.3 THE HISTORY OF EVOLUTIONARY THOUGHT EQUALS ATHEISM

- Many evolutionists who have advanced the theory of evolution are atheists.

Topic question

What is one likely reason that the theory of evolution was proposed by Darwin? **Darwin either did not believe in God or specifically denied His existence. It is impossible to believe that the origin of life rests with a Creator that you do not believe in.**

17.4 EVOLUTION: HOW LIKELY IS IT?

- Biostatisticians have calculated that the chance of life arising in the way evolutionary thought states it did is 1 in 10^{57800}. That last number is a one with 11 full pages of zeros after it. It is a chance so small that it is hard to understand how small a chance it is.

Topic question

Do you think it is possible that life was able to form billions of years ago in the way evolutionists believe it did? Why or why not? **This is an opinion question that I hope you will discuss with your student to understand where they are as far as believing in evolution, have they put any thought into it, etc.**

17.5 CREATION

- Creation is the belief that everything was created for a specific purpose by God. Nothing is an accident.

Topic question

What are the two main thoughts on the origin of life? **Creation and evolution.**

17.6 CREATIONISTS

- People who believe that creation adequately explains the origin of life are called creationists.

Topic question

What is a creationist? **A person who believes that the origin of life is God's creation.**

17.7 EVOLUTION AND CREATION AS PHILOSOPHIES

- At the most basic level, creation and evolution are two different philosophies regarding the origin of life.
- These two different philosophies affect the way in which data is interpreted.
- Despite what is taught in schools and the media, evolution is not fact because it cannot be proven.

Topic question

How do philosophies affect the way we interpret data? **What we believe determines the conclusions we come to regarding information presented to us. If a person's philosophy regarding food is that they do not like vegetables, it is unlikely they will eat a vegetable because they believe they do not like them. If a creationist is presented with information regarding the origin of life, that information is put into a creation context. If an evolutionist is presented with the same information, that information is put into an evolutionary context.**

17.8 THE FOSSIL RECORD

- Fossils are the remains of a previously-living life form.
- An artifact is past evidence specifically of human activity.
- Archaeologists study the past by looking at fossils, artifacts, and other data relating to the past.

Topic question

What is the difference between an artifact and a fossil? **Fossils are the remains of a previously-living life form. An artifact is past evidence specifically of human activity**

17.9 STRATA

- Strata are layers of rock.
- Geologists are scientists who study the earth.

Topic question

What is a geologist? **A scientist who studies the physical processes and land formation of the earth.**

17.10 DO CURRENTLY OBSERVED PROCESSES SUPPORT SLOW OR RAPID FOSSIL FORMATION?

- Evolutionists believe that fossils are formed slowly over time as they are slowly covered by silt in bodies of water.

- Currently observed processes do not support the fact that fossils are formed by being slowly covered by silt over hundreds to thousands of years.

- Creationists believe that rapid fossil formation is a better explanation for how fossils form because that is what is observed today.

Topic questions

Describe the different viewpoints regarding fossil formation. **Evolutionists believe fossils formed slowly over hundreds, thousands, or millions of years. In order for life to have taken five billion years to evolve, slow fossil formation is necessary. Creationists believe that the majority of fossils were formed rapidly, such as would occur during violent flooding conditions of Noah's time.**

What is a polystrate fossil? **It is a fossil that spans many strata; it is not limited to one strata.**

Why are polystrate fossils a problem for evolutionists? **Because a polystrate fossil cannot have formed slowly. It had to have formed by being covered rapidly with dirt and sediment.**

Do sedimentation studies indicate that strata can be formed rapidly? **Yes.**

17.11 DATING TECHNIQUES

- There are various methods to obtain an age of rocks and fossils. They all rely on radioactive decay detection.

Topic questions

If radiometric dating techniques accurately measured the age of rocks and fossils, would the dates obtained by different methods agree with one another? **Yes, they should if the methods are accurate.**

Do they? **No.**

What is the critical part of radiometric dating? **Determining how much of a radioactive isotope a sample originally contained.**

17.12 HALF-LIFE

- The half-life is how long it takes for 50% of the original radioactive isotope in a sample to decay.

- The age of a fossil can be determined by calculating the number of half-lives that have passed for a particular radioactive isotope.

- Two critical assumptions need to be made in order for radiometric dating to give accurate fossil ages. The first is that organisms in the past accumulated radioactive isotopes in *exactly* the same way and amounts as they do today. The other critical assumption is that nothing other than radioactive decay can change the amount of a radioactive isotope in a sample. We will address how valid these assumptions are in the last section of this chapter.

Topic questions

If two half-lives of carbon-14 have passed, how old is a sample? **11,200 years old.**

If two half-lives of potassium-40 have passed, how old is the sample? **2,600,000,000 years old.**

17.13 GEOLOGICAL TIME SCALE

- The geologic time scale is a way in which the age of the earth is broken into subdivisions of Eras, Periods, Epochs, and others.

Topic questions

What is the geologic time scale? **The subdivision of the age of the earth into Eras, Periods, and Epochs, among others.**

How are the subdivisions of the geologic time scale determined? **By the events that supposedly took place during the times indicated.**

17.14 RADIODATING (*OR CARBON DATING*) PITFALLS

- The critical assumption of what the starting amount of radioactive material in a fossil was at the time the fossil was formed may not be accurate.

- The critical assumption that only radioactive decay can lower the amount of a radioactive material in a rock sample may not be true.

- Creationists believe it is likely that both critical assumptions evolutionary radioactive methods use are wrong.

Topic question

What is the evidence that radioactive dating may not accurately reflect the age of fossils/rock? **Organisms may not have accumulated radioactive isotopes in *exactly* the same way and amounts as they do today. If they accumulated less radioactive material during their lifetimes than radio daters think, they would appear older than they actually are when dated. Processes other than radioactive decay may be able to reduce the amount of a radioactive isotope in a sample. If that is true, then it would appear that more of the radioactive material had decayed than actually did. This would lead to the rock being dated older than it actually is. The radiodating methods often do not agree with one another regarding the age of a specimen. If the methods are so accurate and precise as evolutionists claim, there should be little disagreement between the dating methods. However, this is not the case. These are three excellent reasons to doubt the accuracy and the precision of the dating methods.**

17.16 KEY CHAPTER POINTS

- Evolution is the philosophy that both the earth and all life on earth are here due to pure chance alone. Further, all newer life forms are believed to have arisen directly from older life forms. People who believe this to be true are called evolutionists.

- Creation is the philosophy that the earth and life on earth was created by an all-powerful Creator and is here for a reason. People who believe this to be true are called creationists.

- Evolution and creationism are completely opposite explanations for the origin of life. If one completely understands evolution, one cannot believe in creationism, and vice-versa.

- Contrary to what is taught by evolutionary sources, the fossil record does not completely support evolution as "truth." There are many discrepancies in the fossil record that creationists believe should cast serious doubt on evolution.

- Radioactive dating techniques often do not agree with one another regarding the ages of rocks and fossils. This represents another pitfall for evolution.

18 | Evolution and Creation: Principles and Evidence

18.0 CHAPTER PREVIEW

In this chapter we will:

- Review the basic philosophies of evolution and creation.
- Investigate the origins of evolutionary thought and the creation counterpoint.
- Discuss the basic evolutionary principles of descent with modification and natural selection as explanations for the origin of new species and the creation counterpoint.
- Be surprised to find out that creationists believe in natural selection, but do not believe that it can lead to the formation of new species.
- Discuss proposed evolutionary patterns and the creation counterpoint.
- Critically examine evolutionary "evidences" and the creation counterpoint.

18.1 CREATION AND EVOLUTION: A REVIEW

- A naturalists is someone who believes that nature has shaped life and how it has developed over the years. Naturalists are evolutionists.
- Evolutionists believe that all life forms evolved from single-celled organisms.
- Creationists call that particle-to-person or amoeba-to-man evolution.
- The leading evolutionists of the present and past are/were either agnostic or atheists.
- Modern evolutionary thought is a direct result of atheism or agnosticism.

Topic questions

If a person believes that all life forms are on earth due to an accidental arrangement of molecules that formed single-celled organisms, then all life forms directly evolved from that one life form, what is the term used to describe that origin of life philosophy? **Evolution or naturalism.**

Why must atheists develop a different explanation for the origin of life? **It is difficult to believe that the universe was created by God when you specifically deny the existence of God.**

18.2 THE BEGINNINGS OF EVOLUTIONARY THOUGHT

- Acquired characteristics are those traits acquired, or lost, during a person's lifetime. These are traits that are not encoded by DNA (an example of an acquired trait is the loss of finger).
- Scientists such as Lamarck, Blythe, and Wallace provided some foundational thoughts for Darwin.
- Darwin developed the theory of evolution following a long journey on the ship *H.M.S. Beagle.*
- Neo-Darwinism is the current explanation to account for the role of genetics in evolution.

Topic questions

Why can acquired traits not be passed from parent to offspring? **Because acquired traits are not encoded by DNA. Traits that are not controlled by DNA cannot be passed from parent to offspring.**

Who is Charles Darwin? **He is the father of evolution. He wrote the book *The Origin of Species*, which laid the foundation for evolutionary thought.**

What is neo-Darwinism? **The modification of Darwin's theory to account for the contribution that genetics must have for evolution to occur.**

18.3 DESCENT WITH MODIFICATION

- Descent with modification is the first principle of evolution. It states that all organisms on the planet are the modified descendants of their ancestors.

- The natural conclusion of descent with modification is that all organisms on earth are descended from one common ancestor.

- Creationists do not believe that descent with modification can lead to the formation of new life forms.

- Creationists believe that all life forms were created by God to look and act a certain way.

Topic question

What is the difference between evolution and creation if both evolutionists and creationists believe in descent with modification? **Both agree that descent with modification occurs. It is easily observed in nature that traits of multiple generations of organisms can change and those changes are passed on to successive generations. However, creationists disagree with the notion that the changes in traits over time can lead to the formation of an entirely different organism.**

18.4 NATURAL SELECTION

- Natural selection is the biological process that organisms with traits that are beneficial to them are more likely to live than those organisms with traits that are less beneficial. The organisms that survive as a result of their beneficial traits pass the beneficial traits to their offspring. The organisms that do not live do not pass along their unbeneficial traits.

- Natural selection leads to the organism developing adaptations to their environment. Adaptations are traits that are beneficial to the organism in the environment where it lives.

- Evolutionists believe that natural selection and descent with modification leads to evolution of new kinds of organisms (species).

- Creationists believe that natural selection does not lead to the formation of new species; it only leads to a change in the traits of already-created species.

Topic questions

What does it mean to be reproductively isolated? **It means that groups of organisms do not breed with one another.**

If I told you that two types of birds that lived in the same area did not ever breed with one another, would you conclude that they were separate species? **Yes. If they never bred with one another, this would fit the definition that the two types of birds were different species.**

18.5 NATURAL SELECTION: NOT JUST EVOLUTIONARY

- Evolutionists and creationists both believe natural selection occurs because it is an easily-observable process in nature.

Topic question

Both creationists and evolutionists agree that natural selection occurs. True or False? **True.**

18.6 PATTERNS OF EVOLUTION: COEVOLUTION

- Coevolution is the change of two or more species in close association with one another.

Topic questions

A bee pollinating a flower and the flower feeding the bee are examples of coevolution according to evolutionists. True or False? **True.**

Creationists believe the relationship of the bee pollinating the flower and the flower feeding the bee was specifically created by God to work exactly that way. True or False? **True.**

18.7 PATTERNS OF EVOLUTION: CONVERGENT EVOLUTION

- Convergent evolution describes the evolutionary process that causes two completely different species to develop similar structures.

Topic questions

Give a classic example of convergent evolution according to evolutionists. **The presence of wings on insects, birds, and mammals (bats). These species are completely unrelated, but all three evolved the ability to fly. The student may also answer "eyes." These are also listed as examples of convergent evolution by evolutionists.**

How would a creationist explain the presence of wings on these very different organisms? **God knew exactly what the needs of every organism He created was and would ever be. He created them with wings because He knew they would need them. If the student answered eyes for the first question, substitute "eyes" for "wings."**

18.8 PATTERNS OF EVOLUTION: DIVERGENT EVOLUTION

- In divergent evolution, two or more related species become more and more dissimilar over time.

Topic questions

What is the frequent example of divergent evolution evolutionists use to explain divergent evolution? **The Galapagos finches. There are thirteen "species" found on the Galapagos islands. Evolutionists say that all the finch species are descended from one finch species. Through natural selection and adaptive radiation, they became more different from one another and formed into thirteen different species.**

How do creationists explain the same data? **Most creationists agree that the thirteen "species" of Galapagos finches are likely descended from a single common ancestor. However, creationists do not agree that the finches are all different species. Creationists believe that the finches are simply displaying their God-given genetic diversity, which was designed into the DNA at creation.**

18.9 POPULATION GENETICS

- Population genetics is the study of the traits of a specific population and how they change over time.

Topic question

What is genetic drift? **When observed traits of a population change over time.**

18.10 "EVIDENCE" FOR EVOLUTION: TRANSITIONAL FORMS

- A transitional form is an intermediate fossil form that shows a supposed in-between stage as one species evolved to the next.

- According to evolutionary theory, there should be transitional forms found in the fossil record; there are none.

Topic question

Do you think that the *Archaeopteryx* fossil is a bird or a transitional form between a reptile and a bird? **This is an opinion question simply to stimulate thought.**

18.11 "EVIDENCE" FOR EVOLUTION: HOMOLOGOUS STRUCTURES

- A homologous structure is a body part that appears in a wide variety of different types of organisms.

- Evolutionists feel that similar structures indicate the organisms with the similar structures share a common ancestor.

- Creationists believe that homologous structures exist because God knew the best way to design a structure, so He used it in whatever organism needed it.

Topic question

Why don't creationists think that homology indicates evidence for evolution? **Creationists believe that if homology were true, then the genes that code for the similar structures would also be similar, which they are not. Also, creationists think that God knew exactly the best way to design a particular structure; when He found a way that worked best, He used it whenever He needed to.**

18.12 "EVIDENCE" FOR EVOLUTION: EMBRYOLOGY

- Early on in evolutionary theory, a man named Haeckel faked drawings of embryos. He drew them in such a way that embryos of different types of organisms looked almost the same. Evolutionists used to use this as a strong argument that evolution is true until they all found out this was a lie. Unfortunately, there are still modern text books that reproduce Haeckel's drawings, even though they know they are not accurate.

Topic question

18.13 "EVIDENCE" FOR EVOLUTION: BIOCHEMICAL

- Similarities in biochemical processes are used by evolutionists as an argument that evolution is true.

- Creationists believe the similarities in biochemistry are not as close as evolutionists make them out to be. Also, they believe that God used similar biochemistry among different species because that is what worked the best.

Topic questions

If one uses the biochemical process of mutation rates, does the current DNA mutation rate indicate that humans and chimps could have split from a common ancestor ten million years ago? **No. Using the current DNA mutation rate in humans, it would take 352 billion years for humans to have developed enough mutations in their DNA to account for the difference seen in human and chimp DNA sequences.**

What is a major problem with neo-Darwinism? **In order for evolution to occur, organisms need to undergo mutation to their DNA that adds information. There has never been one observed mutation that has added information. All mutations delete information.**

18.14 "EVIDENCE" FOR EVOLUTION: VESTIGIAL STRUCTURES

- Vestigial structures are, by definition, structures that once had a function but now no longer do. Evolutionists use the presence of vestigial structures as proof that ancestors are shared by unrelated organisms.

- Unfortunately, the list of vestigial structures has grown from about 180 in the late 1800s to nearly zero at present. Most current evolutionists do not use this argument anymore, but a lot of text books still do.

Topic question

18.15 CONCLUSION

- Do not learn to blindly believe things. Develop a healthy skepticism and learn for yourself what the facts do and do not support.

18.17 KEY CHAPTER POINTS

- Evolution is a way to explain the origin of life and how it appears today without the existence of a Creator.

- Creationists believe God created the universe and all life on earth.

- Charles Darwin is considered the "father" of modern evolutionary thought based on his book *The Origin of Species,* published in 1859.

- The evolutionary processes of descent with modification and natural selection are thought by evolutionists to give rise to new species. This process occurs slowly.

- Creationists believe all species were created by God.

- Evolutionists think several patterns of evolution (including coevolution, convergent evolution, and divergent evolution) explain the presence of life on earth by descent with modification and natural selection.

- Creationists believe natural selection does occur, but that it does not lead to the formation of new species; rather, all species are created with much genetic potential to adapt as a result of natural selection.

- Evolutionists think the studies of transitional forms, homologous structures, embryology, biochemistry, and vestigial structures contribute to the evidence that evolution is true.

- Creationists believe these studies simply indicate the intelligent design of the Creator; He used the design that worked best and stuck with it.

19 | Speciation and Adaptation

19.0 CHAPTER PREVIEW

In this chapter we will:

- Discuss the concept of neo-Darwinism and its validity in explaining the formation of new species.
- Investigate the fact that meaningful gene-producing mutations have never been observed to occur, casting doubt on the concept of neo-Darwinism.
- Discuss adaptations and the processes that evolutionists and creationists think explain them.
- Investigate natural selection in action on the Galapagos Islands and explain why it is not an example of evolution in action.
- Discuss the concepts of gradualism and punctuated equilibrium.
- Study the evolution of man.
- Wrap up evolution and creation.

19.1 OVERVIEW

- Neo-Darwinism attempts to explain evolutionary principles from a genetic standpoint.
- Creationists do not believe this offers a valid explanation for evolution.

Topic question

19.2 NEO-DARWINISM AND EVOLUTION

- According to neo-Darwinian theory, organisms evolved because they acquired new genes which coded for new traits, causing one species to evolve into another.

Topic question

Why is it that, according to neo-Darwinism, in order for evolution to be true, species need to gain genetic material (genes)? **According to evolution, less complex species evolve into more complex species. This cannot happen unless the less-complex species gains new genes coding for the more-complex traits of the new species.**

19.3 NEO-DARWINISM: SEEMS LOGICAL....CAN IT HAPPEN?

- Unfortunately, geneticists have never been able to identify or create one gene-causing mutation.
- Since information-adding mutations do not occur, how can neo-Darwinism explain the formation of new species?

Topic question

What is the scientific problem preventing neo-Darwinism from explaining the formation of new species? **Mutations do not add information, they only delete it. Since there is no known scientific way for information-adding mutations to occur, neo-Darwinism cannot possibly explain how a less-complex species evolves into a more-complex one.**

19.4 ADAPTATIONS

- Adaptations are traits that enhance a species' survival in its environment.
- There are structural, behavioral, and physiological adaptations.

- Evolutionists believe that adaptations arise through natural selection working on new traits caused by mutation.

- Creationists believe that God gave all species whatever amount of genetic and trait variability they needed to form adaptations. Only God knows exactly what a species will need.

Topic questions

What is the difference between the evolutionary and creation viewpoint on adaptations? **Evolutionists believe that random mutations just happen to add the necessary information needed for a species to develop new and specialized adaptations. Creationists believe adaptations are the result of creation. God created all species with exactly the necessary genetic information to develop adaptations.**

Does current genetic knowledge support the evolutionist's viewpoint? **No. There are no information-adding mutations known to occur that would enable a species to develop a new trait.**

19.5 VARIATIONS

- A beneficial variation of a trait of a species is selected for through natural selection.

- An unbeneficial trait variation is selected against through natural selection.

Topic question

Why is trait variation necessary for a species to survive? **The only way for natural selection to occur is through trait variation. If all organisms of a species were *exactly the same*, then no trait variation would exist. When trait variation is reduced, natural selection has nothing to work on. Reduced trait variation is one of the things that can lead a species to become extinct. Natural selection works on these trait variations to eventually produce a new adaptation.**

19.6 VARIATION AND THE GALAPAGOS FINCHES: EVOLUTION OR NATURAL SELECTION?

- The Galapagos finch study by Dr. Grant is an excellent example of natural selection in action.

Topic question

The Galapagos finch study by Dr. Grant is an excellent example of evolution in action. True or False? **False.** Why did you answer as you did? **The change in the Galapagos finch population as a result of drought and rainfall cannot be evolution because there has been no change in the genetic information the bird population contains. In order for evolution to occur, there has to be a gain in genes and a new species formed. All that was seen in the Galapagos finch study was natural selection acting on trait variation.**

19.7 CREATIONISM AND THE GALAPAGOS FINCHES

- Creationists believe that God gave the ancestor Galapagos finch great genetic variability so that it could respond to natural selection and live in a variety of environments on the Galapagos.

Topic question

Do creationists believe the Galapagos finches all share the same ancestor finch? **Yes. Creationists believe that the Galapagos finches were created with great genetic and trait variability on which natural selection could act.**

19.8 GRADUALISM

- Gradualism is the belief that all species evolved slowly over a long period of time.

- Unfortunately the fossil record does not support this theory.

Topic question

Why do creationists think that gradualism is not a reasonable way to explain the origins of species? **Because if gradualism were true, then there should be a fossil record full of** transitional fossil forms. There are none. In addition, the appearance of species in the fossil record shows the abrupt appearance of fully-formed species. The species that are in the fossil record and are still alive today have not changed at all. This goes against gradualism and evolution. Finally, there is at least one mass extinction in the fossil record. Evolutionists believe there are more, but that is based on the possibly faulty logic that strata form slowly and fossils take thousands or millions of years to form. Evolutionists do not know what causes mass extinction. Creationists do—they say mass extinction was caused by the Great Flood of Noah.

19.9 PUNCTUATED EQUILIBRIUM

- Punctuated equilibrium tries to explain the long periods of stasis and rapid explosions of new and fully-developed life forms with no intermediates from an evolutionary point of view. It does such a poor job of it that one of the scientists who proposed it backed off of the theory later in his life.

Topic question

What is punctuated equilibrium and why is it a weak attempt to explain what the fossil record shows in regards to the abrupt appearance of fully-formed species with no intermediates? **The theory of punctuated equilibrium tries to explain the long periods of stasis and rapid explosions of new and fully-developed life forms with no intermediates from an evolutionary point of view. It explains the stasis and rapid change by natural selection slowly preparing the organism for a species change or mass adaptation change (the "equilibrium" part). It is as if nature "knows" the organism needs to undergo trait changes. Then all of a sudden, the traits explode out and new species are formed (the "punctuated" part). But what are all those genes doing just sitting around in an organism? Natural selection says that if a gene codes for a trait that has no benefit for an organism, it will be selected against. If a gene is sitting around in a species and does not code for a beneficial trait, the gene should be selected against. But if the genes were able to just sit in the nucleus and do nothing, waiting to be expressed so the new species can appear, how do the genes know when they should turn on so the new species can develop?**

19.10 HUMAN EVOLUTION

- Humans are classified as primates because we share certain trait with primates.

- The only conclusive evidence in the fossil record indicates that humans did not originate from anything but humans.

Topic question

According to the fossil record, what species is the closest ancestor to humans? **Chimpanzees.**

19.11 EVOLUTION CONCLUSION

- The only objective conclusion that can be made regarding evolution is that there is no scientific evidence that supports it. As such, those who believe evolution is true believe in the religion of evolution. They believe it on faith, because evolution is based on assumptions that cannot be proven.

- Creationism, although also founded on faith, objectively does a better job of explaining data regarding the origins of species than evolution.

Topic question

What do you believe explains the presence of life, evolution or creation? Know why you answer as you do!!!

19.13 KEY CHAPTER POINTS

- Neo-Darwinism is the evolutionary attempt to explain evolution on a genetic basis.

- In order for neo-Darwinism to properly explain evolution, some as-yet-unknown mechanism must be present allowing genetic mutations to add meaningful genetic information that codes for new traits. Current genetic knowledge indicates mutations do not add information, they delete it.

- Attempts to understand the length of time it would require information-adding mutations to occur to cause one species to evolve into another indicate that it would take much longer for evolution to occur than evolutionists say it does.

- Organisms have various types of adaptations that allow them to function more effectively in their environments.

- Evolutionists believe adaptations are a result of information-adding mutations. Creationists believe adaptations are the result of natural selection working through genetic diversity endowed on the organism when it was created.

- Changes in the Galapagos finch population as a result of natural selection are an example of natural selection and not evolution because there has been no change in the genetic information of the birds.

- Evolutionists believe that speciation occurs gradually. This is called gradualism.

- Evolutionists believe there have been many mass extinctions. Creationists believe there has been one.

- Punctuated equilibrium is the evolutionary attempt to explain what the fossil record documents—species simply "appear" in their fully "evolved" state.

Biological Classification 20 and Viruses

20.0 CHAPTER PREVIEW

In this chapter we will:

- Review a brief history of classification schemes.
- Discuss the binomial classification system devised by Linnaeus, which is still in use today.
- Introduce the six classification kingdoms of Archaebacteria, Eubacteria, Protista, Fungi, Plantae, and Animalia.
- Investigate the structure and function of the nonliving particles called viruses.

20.1 OVERVIEW

- Living organisms have been classified for more than two thousand years.

Topic question

Why was the classification system of Aristotle abandoned after nearly sixteen hundred years? **Aristotle's system relied on using common names of animals and plants. Many organisms are known by more than one common name. This resulted in great confusion because common names are not specific enough, so a new system was needed to decrease confusion and increase specificity of naming organisms.**

20.2 LINNAEUS

- Carol Linnaeus developed the modern method of organism classification in the sixteenth century.

Topic question

Who invented the modern system of classification? **Carolus Linnaeus.**

20.3 TAXONOMY

- Taxonomy is the science of classifying organisms.
- There are seven levels of order in taxonomy.

Topic question

What are the seven levels of order in the modern classification system? List them in order from least specific to most specific. **Kingdom, Phylum, Class, Order, Family, Genus, and Species.**

20.4 BINOMIAL NAMES

- The binomial system is another name used for the modern classification system.
- All organisms are assigned a two-part name, and only one specific organism type has a given binomial name.

Topic questions

If an organism's binomial name is *Bacillus anthrasis*, what is its genus? **Bacillus.**

What is its species? **Anthrasis.**

20.5 METHOD OF CLASSIFICATION

- Organisms are classified based on common similarities.

- Similarities which are used for classification are structural and biochemical/physiological.

- Characteristics used for classification are prokaryote/eukaryote cell structure,

Topic questions

What is a structural similarity? **A similarity in the way two or more organisms look; common shared characteristics.**

What is a biochemical or physiological similarity? **A similarity in the way two or more organisms function.**

20.6 PROKARYOTE KINGDOMS

- There are two kingdoms that contain prokaryotes: Archaebacteria and Eubacteria.

Topic question

If you are studying prokaryotic species, into which kingdom(s) could they be classified? **All prokaryotes are classified into Archaebacteria or Eubacteria.**

20.7 Eukaryote Kingdoms: Protista

- Protists are single-celled or grouped organisms that are eukaryotes.

Topic question

What are the defining features of organisms from Protista? **They are unicellular and eukaryotic.**

20.8 KINGDOMS: EUKARYOTE—FUNGI

- The defining characteristics of Fungi are cell walls made of chitin; they are also decomposers.

Topic question

What are the defining characteristics of the organisms from Fungi? **They all have cell walls made of chitin and are decomposers.**

20.9 KINGDOMS: EUKARYOTE—PLANTAE

- The defining characteristics of Plantae are multicellularity, photosynthesis, and cell walls made of cellulose.

Topic questions

What are the defining characteristics of Plantae? **They all have cell walls made of cellulose, perform photosynthesis, and are multicellular.**

You have just picked an organism growing on the ground in the forest and your microscope eyes allow you to see that it has cell walls, but you cannot tell what they are made of. From what kingdom(s) could this organism be? **Fungi or Plantae.**

20.10 KINGDOMS: EUKARYOTE—ANIMALIA

- The defining characteristics of almost all organisms in Animalia are that they have multicellularity, are heterotrophic, have no cell walls, perform sexual reproduction, can move, and have a symmetric body plan.

Topic question

What are the defining characteristics of the organisms of Animalia? **Multicellularity, heterotrophic, no cell walls, sexual reproduction, can move, and have a symmetric body plan.**

20.11 VIRUSES

- A virus is not a living organism. It is called a particle by biologists.

Topic questions

Why are viruses not considered "alive"? **Viruses do not exhibit several properties that all true forms of life exhibit. Viruses do not grow, do not independently reproduce, do not perform any of their own metabolic processes, and cannot maintain homeostasis.**

Which nucleic acids can viruses contain? **RNA or DNA.**

What does virulent mean? **It means "disease causing."**

Why are retroviruses important in recombinant DNA? **Retroviruses contain reverse transcriptase, a critical enzyme in some recombinant technology.**

20.12 VIROIDS AND PRIONS

- Viroids are the smallest-known particles able to direct their own replication.

- Viroids are small protein pieces that clump inside of cells and kill them.

- Prions are infectious protein particles that commonly clump in brain cells and cause memory loss and death.

Topic questions

What does a viroid contain that makes it infectious? **RNA.**

What does a prion contain that makes in infectious? **A protein.**

What about prions make them unique? **They are able to reproduce themselves without a nucleic acid. The protein a prion contains is somehow able to direct its own reproduction.**

20.14 KEY CHAPTER POINTS

- Biological classification is the systematic placement of all living organisms into categories based on their anatomy (the way they look) and their biochemistry (the way they function).

- The modern classification system is a seven-level system in which all organisms are first placed into an appropriate kingdom. They are then placed into a phylum, class, order, family, genus, and species.

- All organisms are referred to by scientists using their binomial name (*Genus species*). This avoids confusion since only one organism has a given binomial name.

- The six kingdoms are Archaebacteria, Eubacteria, Protista, Fungi, Plantae, and Animalia.

- Viruses, viroids, and prions are not technically considered "life" because they are unable to reproduce themselves independently.

- Viruses have been studied well due to their ability to cause disease in plants, animals, and humans. They have well-defined life cycles.

21 | Kingdom Eubacteria and Archaebacteria

21.0 CHAPTER PREVIEW

In this chapter we will:

- Understand the main differences between the organisms of Eubacteria and Archaebacteria.
- Discuss common bacterial cell shapes and review bacterial cell structure.
- Investigate Eubacterial cell wall structure and how it relates to the Gram staining procedure.
- Discuss the mechanism of action of antibiotics.
- Review bacterial modes of obtaining energy and manufacturing organic molecules.

21.1 OVERVIEW

- Bacteria are the organisms with the simplest structure that can independently carry out the functions of life.
- Bacteria are divided into the Eubacteria and Archaebacteria Kingdoms.

Topic questions

What is a bacteria? **It is the organism with the simplest physical structure that can sustain the activities of true life.**

What two kingdoms are bacteria placed? **Archaebacteria and Eubacteria.**

21.2 GENERAL BACTERIAL PROPERTIES

- Bacteria are microscopic and ubiquitous (found everywhere).
- Bacteria are both helpful and harmful to humans.

Topic questions

Name some helpful uses of bacteria. **They can be used to make yogurt, cheese, and buttermilk. They are also the final pathway for dead and decaying matter to return to a useable form. Bacteria break down sewage before it is released into the water supply. They help break down food in the intestines of animals and humans and also protect against infections.**

Are all bacteria helpful to humans? **No, some of them cause minor and serious infections.**

21.3 BACTERIAL SHAPES

- Bacteria are found in three common shapes—rods (bacilli), spheres (cocci), and spiral (spirochetes).

Topic question

What are the three common bacterial shapes? **Rod (bacillus), sphere (cocci), and spiral (spirochete).**

21.4 BACTERIAL STRUCTURE

- No bacterial cells have a nucleus.
- DNA is kept in an area of the cell called the nucleoid.
- All bacteria have a cell wall.

- Some bacteria have a capsule.
- Photosynthetic bacteria do not contain chloroplasts.

Topic questions

If you are standing inside of a bacterium and are surrounded by DNA, in what area of the cell are you standing? **The nucleoid.**

Let's say this bacterium has a capsule. If you decide to walk out of the bacteria, as you walk out of the cell, in which order will you walk through the capsule, cell membrane, and cell wall? **The cell membrane is first, then the cell wall, then the capsule.**

If photosynthetic bacteria do not contain chloroplasts, how can they perform photosynthesis? **Bacteria contain chlorophyll and other enzymes in their cytoplasm that perform photosynthesis.**

21.5 ENDOSPORES

- An endospore is a bacterial structure that forms in times of environmental stress.

Topic questions

All bacteria produce endospores. True or False? **False. Some bacteria form endospores, not all.**

What is contained in an endospore? **DNA (nuclear material) and cytoplasm surrounded by an inner and an outer spore coat.**

21.6 BACTERIAL CELL WALLS

- All bacteria have cell walls.
- Eubacterial cell walls are made out of peptidoglycan.
- Archaebacterial cell walls are made from other types of polysaccharides and proteins.

Topic questions

What molecules make up Eubacterial cell walls? **Peptidoglycan.**

All bacteria have cell walls. True or False? **True.**

21.7 GRAM STAIN

- A Gram stain is a useful staining procedure to identify whether a bacterial cell is Gram-positive or Gram-negative.

Topic questions

Why does a Gram-positive bacterial species stain Gram-positive? **Gram-positive bacteria have much thicker peptidoglycan cell wall layers. When the first stain of the Gram stain (crystal violet) is added, the thick peptidoglycan layer retains the dye throughout the Gram stain procedure. Since the dye is not washed out during the washing steps, the bacteria is said to be Gram-positive. Gram-positive bacteria look blue under the microscope after they have been stained.**

Why does a Gram-negative bacterium stain negatively? **The cell wall of a Gram-negative bacterium has a thin peptidoglycan layer. It does not retain the crystal violet during the washing steps of the Gram stain. The final step of the Gram stain is the application of a pink colored dye called safranin. Gram-negative bacteria stain with this stain and look pink under the microscope.**

21.8 OTHER BACTERIAL STRUCTURES: CAPSULE, PILLI, AND FLAGELLA

- Some bacteria have an outer polysaccharide layer called a capsule. Some capsules have a glycocalyx.
- Pilli are small proteins projecting from the surface of a bacterium.
- Some bacteria move through the actions of flagella.

Topic questions

What is a capsule? **A polysaccharide coating some bacterial species have.**

Are prokaryotic and eukaryotic flagella the same? Why or why not? **They are not the same. They are both made out of proteins, but not the same proteins. In addition, bacterial flagella are much smaller than eukaryotic flagella.**

21.9 EUBACTERIA

- Eubacteria are Gram-positive cocci, Gram-positive bacilli, Gram-negative cocci, Gram-negative bacilli, coccobacilli, and spirochetes.

Topic questions

What are endotoxins and which type of bacteria produces them? **They are a mixture of lipids and carbohydrates formed from the outer membrane when a Gram-negative bacterium breaks apart.**

What are exotoxins and what type of bacteria produces them? **They are proteins secreted by some Gram-positive bacteria.**

What is a coccobacillus? **It is a bacterial shape that is kind of like an elongated sphere or a squished up bacillus.**

21.10 HOW DO ANTIBIOTICS WORK?

- Antibiotics work to inhibit a bacterial function that the host does not perform. In that way, they are able to kill bacteria and not harm the host.

Topic questions

How do antibiotics like penicillin and cephalosporin work? **They inhibit a bacteria's ability to make the cell wall.**

How does antibiotic resistance occur? **Some bacteria become resistant by acquiring plasmids, which contain enzymes to break the antibiotic down before it has a chance to kill the bacteria. Other bacteria become resistant through natural selection.**

21.11 MODES OF OBTAINING ENERGY AND CARBON

- Bacteria exhibit diverse ways of obtaining energy and carbon.
- Bacteria can be photoautotrophs, chemoautotrophs, and chemoheterotrophs.

Topic questions

What is an obligate anaerobe? **They are bacteria that cannot live in the presence of oxygen.**

What is a saprophyte? **It is an organism that feeds off of dead organic matter.**

Some bacteria show symbiotic relationship with other organisms. What does this mean? **A symbiotic relationship is usually defined as a relationship between two organisms in which the presence of each organism benefits the other and no harm is done to either one.**

21.12 ARCHAEBACTERIA

- Archaebacteria are called extremophiles because they often live in very harsh (extreme) environments.

Topic questions

Why are the organisms of Eubacteria and Archaebacteria separated into two kingdoms? **Because the cell walls of the Eubacteria and Archaebacteria are structurally different. Also, the sequence of bases in the tRNA and rRNA of the two kingdoms is different.**

What is a methanogen? **It is an Archaebacteria that produces methane during carbon fixation.**

21.14 KEY CHAPTER POINTS

- Prokaryotes are mainly bacteria and are classified into the Kingdoms of Archaebacteria or Eubacteria.

- Bacteria are the most numerous organisms on earth and live in a wide variety of locations and environmental conditions.

- Bacteria have three common shapes: bacillus (rod shaped), coccus (spherical), and spiral.

- Bacteria have a cell wall surrounding the cell membrane as an added layer of protection.

- Some bacteria have an additional layer outside the cell wall called the capsule.

- There are two types of cell walls that stain differently during the Gram stain procedure.

- Many Eubacterial species infect and cause disease in humans and animals.

- Antibiotics work against bacteria, because bacteria perform certain biochemical processes that humans and animals do not. Antibiotics prevent bacteria from performing critical processes, which kills the bacteria.

- Bacteria have the most diverse means of obtaining their energy molecules.

- Archaebacteria are adapted to live in harsh environments and are called "extremophiles."

22 | Kingdom Protista

22.0 CHAPTER PREVIEW

In this chapter we will:

- Discuss how the organisms in Protista acquire their energy source.
- Describe the common body structures of algae.
- Discuss the reproductive cycle of algae, protozoa, and slime/water molds.
- Introduce the concept of the reproductive cycle of alternation of generations, which will be a common mode of reproduction for many Fungi and Plantae species.
- Discuss the attributes of organisms in Protista.

22.1 OVERVIEW

- Protists are the least complex eukaryotes from a structural and functional standpoint.
- Protists are composed of the protozoa, algae, slime molds and water molds.
- Protists make up a significant part of plankton.

Topic questions

How are the protists often broken down? **They are broken down into three categories: the animal-like protists, or protozoa—also called heterotrophic protists; the plant-like protists, or algae—also called the autotrophic protists; the fungus-like protists, or slime molds and water molds—also called the absorptive protists.**

What part of the plankton are algae? **Phytoplankton.** What part of the plankton are the protozoa? **Zooplankton.**

22.2 ALGAE: GENERAL

- Algae are diverse.
- Algae contain chlorophylls a and often b, c, and d.
- There are commonly seven phyla of algae.

Topic questions

Why are algae no longer classified with plants? **Plants form their gametes in gametangia (chambers), which are multicellular; algae form their gametes in gametangia, which are unicellular.**

How many phyla of algae are there? **Seven.**

22.3 ALGAE: STRUCTURE

- Algae can be unicellular or multicellular.
- Multicellular algae have a common structure.

Topic questions

What is a stipe? **It is the stem-like part of a multicellular alga.**

Are most forms of algae multicellular or unicellular? **Unicellular.**

Which two types of algae can have a holdfast? **All multicellular algae and some filamentous algae have holdfasts.**

What molecule do the most of the phyla of algae use for their cell wall? **Cellulose. If the student answers carbohydrate, that would be okay, too.**

22.4 ALGAE: PHYLA

- The traits of the seven phyla of algae are briefly discussed.

Topic questions

What is bioluminescence? **The ability of living things to emit their own light.**

Which phylum (or phyla) exhibit bioluminescence? **Dinoflagellata.**

How does *Euglena* move in its environment? **A flagellum.**

22.5 ALGAE: REPRODUCTIVE

- Algae can reproduce sexually and asexually.
- The male unicellular gametangium is called the antheridium, and the female unicellular gametangium is called the oogonium.

Topic questions

What is an algal zygospore? **It is a hearty diploid reproductive structure that is formed in times of environmental stress. The key in this answer is that the structure is diploid and it is a reproductive structure.**

What is alternation of generations? **It is the description of the life cycle in which the organism exists in the haploid state in one generation and in the diploid states in the following generation.**

22.6 PROTOZOA

- Protozoa are the animal-like, or heterotrophic, protists.
- Most have a specialized area called a food vacuole in which they break down their nutrients.

Topic question

What is a food vacuole? **It is a chamber in which the protozoa digest their nutrients.**

22.7 PROTOZOA: PHYLUM SARCODINA

- Sarcodines are characterized by organisms from the genus *Amoeba*.
- Some members of this phylum cause human disease.

Topic questions

How do amoeba move? **Through cytoplasmic streaming. The cytoplasm moves inside the amoeba, which causes the surface of the cell membrane to move in different directions. As the cytoplasm and cell membrane move, it pulls the amoeba.**

What is a pseudopod? **It is an extension of the cell membrane; it forms as a result of cytoplasmic streaming.**

22.8 PROTOZOA: PHYLUM CILIOPHORA

- Organisms form Ciliophora are characterized by organisms from the genus *Paramecium*.

Topic questions

How do organisms form Ciliophora move? **Cilia.**

When a paramecium performs conjugation, is the macronucleus or micronucleus exchanged? **The micronucleus.**

22.9 PROTOZOA: PHYLUM ZOOMASTIGINA

- Zoomastigina includes organisms that live in fresh water lakes and streams and can cause human infections.

Topic question

What is Leishmaniasis? **It is a disease cause by *Leishmania*, an organism from the phylum Zoomastigina, which causes leishmaniasis.**

22.10 PROTOZOA: PHYLUM SPOROZOA

- Sporozoans cause the most human disease of any other Phylum of organism.
- Malaria is caused by a sporozoan from the genus *Plasmodium.*

Topic questions

What causes malaria? **There are several organisms from the genus *Plasmodium* that cause malaria.**

What is the vector for *Plasmodium*? **The *Anopheles* mosquito.**

What is the host? **Humans.**

What two forms of *Plasmodium* live inside of the vector? **The sporozoite and gametocyte.**

What two forms live inside the host? **The sporozoite, merozoite, and the gametocyte.**

22.11 SLIME MOLDS AND WATER MOLDS: THE FUNGUS-LIKE PROTISTS

- Slime molds are placed in the phyla Acrasiomycota and Myxomycota.
- Water molds are in the phylum Heterokontophyta.

Topic question

What are the three phyla into which the water and slime molds are placed? **Acrasiomycota, Myxomycota, and Heterokontophyta.**

22.12 SLIME MOLDS: ACRASIOMYCOTA AND MYXOMYCOTA

- Slime molds are found in wet places like forest floors.
- When the slime molds go into the reproductive phase, they move into a warm and dry place.

Topic question

What do the slime molds do when it is time to reproduce? **They move to a warm and dry place and produce fruiting bodies.**

22.13 WATER MOLDS: HETEROKONTOPHYTA

- Water molds are important, disease-causing organisms.

Topic question

What was responsible for the Irish famine in the late 1800s? **A water mold infection of the potato, called potato blight. It is caused by a water mold from the phylum Oomycota.**

22.15 KEY CHAPTER POINTS

- Organisms from the Kingdom Protista are mostly unicellular, but there are some multicellular species. They are composed of the protozoa, algae, slime molds, and water molds.
- Protozoa are the "animal-like protists." They are the heterotrophic protists.
- Algae are the "plant-like protists." They are autotrophic protists.
- Slime molds and water molds are the "fungus-like protists." They are the absorptive protists.
- Most all protists are able to reproduce sexually and asexually.
- Algae are photosynthetic. The species are diverse—some being microscopic, while others are quite large.
- The protozoa are mainly heterotrophic. Amoeba and paramecium are the prototypical protozoans.

- Species from the protozoa phyla of Zoomastigina and Sporozoa commonly cause human disease. Malaria is caused by a Sporozoan.

- Slime molds and water molds are absorptive organisms. They are important decomposers in wet environments. Most are multicellular or large, unicellular organisms.

23 | Kingdom Fungi

23.0 CHAPTER PREVIEW

In this chapter we will:

- Discuss the common characteristics Fungi share.
- Define the general structure plan of Fungi.
- Investigate the modes of sexual and asexual reproduction of the fungal divisions of Basidiomycota, Ascomycota, Zygomycota, and Deuteromycota.
- Define the specific structural plans for the four Fungi divisions.
- Investigate the structure and function of lichens.
- Learn the beneficial and harmful features of Fungi.

23.1 OVERVIEW

- Fungi are structurally more complex than protists.

Topic question

About how many organisms are classified as Fungi? **More than 100,000.**

23.2 FUNGI CHARACTERISTICS

- Fungi are all saprophytes.
- The cell walls of Fungi are made of chitin.
- Fungi reproduce through spores.

Topic questions

How do Fungi obtain their nutrition? **Fungi are saprophytes. They perform extracellular digestion, then absorb the nutrients.**

Are all Fungi decomposers (saprophytes)? **No, there are some species that are important parasites.**

23.3 STRUCTURE

- The basic building block of a fungus is the hypha.
- Hyphae can be septated or non-septated.
- The growing part of Fungi is called a mycelium.

Topic questions

What is a non-septated hyphus called? **A coenocytic hyphus.**

What is a mycelium? **It is the growing part of a Fungus.**

23.4 REPRODUCTION

- Most Fungi exist in the haploid state and can reproduce sexually or asexually.
- Fungi are not male and female, but plus and minus.

Topic questions

Are Fungi usually haploid or diploid? **Haploid.**

What is a spore? **It is a gamete formed through mitosis.**

What is usually the trigger for a fungus to enter into sexual reproduction? **Unfavorable growing conditions.**

23.5 DIVISIONS OF THE KINGDOM FUNGI

- Fungi are divided into four divisions (phyla)—Basidiomycota, Zygomycota, Ascomycota, and Deuteromycota.

Topic question

Into how many divisions (phyla) are the Fungi classified? **Four.**

23.6 DIVISIONS OF THE KINGDOM FUNGI: BASIDIOMYCOTA

- Basidiomycetes are also called the club fungi.
- Mushrooms are the typical basidiomycete.

Topic questions

What is a cap? **It is the top of a basidiocarp (mushroom) and contains the reproductive structures.**

What are basidiospores? **They are the sexual reproductive gametes of basidiomycetes.**

What are gills? **They are slits on the undersurface of the cap where the basidiospores are formed.**

23.7 DIVISIONS OF THE KINGDOM FUNGI: ZYGOMYCOTA

- This division (phylum) is characterized by black bread mold.
- Most zygomycetes live in moist environments.

Topic questions

What are rhizoids and stolons? **Rhizoids are hyphae that grow into the surface of whatever a fungus grows on. Stolons are hyphae that grow over the surface of whatever the fungus is living on.**

What is a zygospore and where does it form? **A zygospore forms at the site where a plus and minus mating type fuse. A zygospore produces haploid spores. When conditions are favorable, a sporangiophore grows out of the zygospore and release the haploid spores.**

23.8 DIVISIONS OF THE KINGDOM FUNGI: ASCOMYCOTA

- Ascomycetes are also called sac fungi.
- An ascocarp forms from the union of an ascogonium and antheridium.

Topic question

What is an ascocarp and how does it form? **An ascocarp is formed from the union of a male (antheridium) and female (ascogonium) reproductive structures. The ascocarp produces fungal spores.**

23.9 DIVISIONS OF THE KINGDOM FUNGI: DEUTEROMYCOTA

- This division (phylum) is also called fungi imperfecti.

Topic question

Why are the deuteromycetes called the fungi imperfecti? **Because they are not known to have any sexual reproductive cycle.**

23.10 YEASTS

- A yeast is a fungus in the unicellular state.

Topic questions

What is a yeast? **A yeast is a fungus that is living in a unicellular state.**

What is a filamentous mycelium? **It is a yeast that has budded and formed many yeast cells, which remain attached to one another.**

23.11 LICHENS

- Lichens are a symbiotic relationship between a fungus and a photosynthetic organism living in its mycelium.

Topic questions

What are the two Fungi divisions that make up the mycelium structure of a lichen? **Either Ascomycota are Basidiomycota.**

Which are the photosynthetic organisms that can be part of a lichen? **Algae or photosynthetic bacteria.**

23.12 MYCORRHIZAE

- Mycorrhizae are the symbiotic relationship that many fungi have with plant roots.

Topic question

What are mycorrhizae? **The symbiotic relationship of a fungus with plant roots.**

23.13 PRACTICAL ASPECTS REGARDING FUNGI

- Fungi are one of the primary decomposers on earth.
- Many species of Fungi cause disease in humans, animals, and plants.
- Many species of Fungi are useful to humans.

Topic questions

How are Fungi harmful to man or plants? **They can cause disease.**

How can Fungi be helpful to man? **Some produce helpful chemicals, like antibiotics. Yeast is used to make bread; mushrooms are eaten.**

23.15 KEY CHAPTER POINTS

- Fungi are more complex structurally than Protists.
- Fungi are all saprophytes, and their cell walls are made of chitin.
- The basic building block of a fungus is the hypha.
- Most fungi are haploid and produce reproductive structures called spores.
- There are four Divisions (Phyla) of Fungi—Basidiomycota, Zygomycota, Ascomycota, and Deuteromycota.
- Basidiomycota are called the club fungi. The typical basidiomycete is the mushroom.
- The typical zygomycete is common bread mold.
- Ascomycetes are also called the sac fungi. Bread yeast is a typical ascomycete.
- Deuteromycetes are also called the fungi imperfecti.
- Lichens are a symbiotic relationship between a fungus and an alga or photosynthetic bacteria.
- Mycorrhizae is the term used to describe the association of a fungus growing in and around plant roots.
- Many fungi cause human and plant diseases.

24 Plants: Introduction, Structure, and Function

24.0 CHAPTER PREVIEW

In this chapter we will:

- Review plant cell structure.
- Discuss the structure of plant cell walls.
- Define the difference between vascular and nonvascular plants.
- Introduce the basic concepts of plant reproduction with special attention to monocots and dicots.
- Investigate the specialized structure and function of the tissues that form the roots, stems, and leaves of plants.
- Discuss the important place that water has in maintaining the structure and function of plants.

24.1 INTRODUCTION TO PLANTS

- Plants are the primary producers on earth.
- Botany is the study of plants.
- All plants are autotrophic via photosynthesis.

Topic questions

What does a botanist study? **Botany or plants.**

Not all plants are autotrophic. True or False? **False. One of the defining features of plants is that they are all photosynthetic, which means they are all autotrophic.**

24.2 PLANT CELL STRUCTURE

- All plants are eukaryotic.

Topic questions

What is one organelle that plant cells contain but animal cells do not? **Chloroplasts. If the student answered cell walls, that is incorrect because a cell wall is not an organelle.**

What is one organelle that plant cells do not have but animal cells do? **Centrioles.**

24.3 PLANT CELL WALL: GENERAL

- Plant cell walls are made of cellulose.
- Cell walls add support and protection for the plant cell.

Topic question

Why is the cell wall porous? **To allow the passage of gasses into and out of the cell and also the movement of nutrients into and wastes out of the cell.**

24.4 PRIMARY CELL WALL

- In plants with more than one cell wall, the first cell wall made is the primary cell wall.
- Primary cell walls are held together by the middle lamella.

Topic questions

What is the primary cell wall? **The first cell wall made by a plant cell.**

What is the middle lamella made of and what is its function? **It is made of pectins, which are sticky and hold adjacent primary cell walls together.**

24.5 SECONDARY CELL WALL

- The secondary cell wall forms between the cell membrane and the primary cell wall.
- The secondary cell wall contains lignin.

Topic questions

What makes up the secondary cell wall? **Cellulose and lignins.**

Which cell wall is stronger, the primary or the secondary? **The secondary.**

24.6 PLASMODESMATA

- Plasmodesmata are small channels that connect plant cells to one another.

Topic question

What are plasmodesmata? **They are small channels that connect plant cells together.**

24.7 VASCULAR AND NONVASCULAR PLANTS

- Some plants have tubes that run throughout the entire plant to carry nutrient and water. These are called the vascular plants.
- Other plants do not have these tubes and are called nonvascular plants.
- The tissues of the vascular plants are called xylem and phloem.

Topic questions

What is the difference between a vascular and a nonvascular plant? **Vascular plants contain tubes that transport materials throughout the plant, nonvascular plants don't.**

What is xylem and which way does it flow? **Xylem is a vascular tissue of plants. It carries water and nutrients from the roots to the rest of the plant.**

What is phloem and which way does it flow? **Phloem is a vascular tissue of plants. It carries glucose from the leaves to the rest of the plant.**

24.8 GENERAL PLANT REPRODUCTION

- Plants can reproduce with seeds, spores, both, or neither. The mode of reproduction depends on the plant.
- A seed and a spore are different structurally.

Topic question

How are a seed and a spore different? **A spore contains only one cell with the ability to grow into a new plant. Spores also do not have any type of protective coating, as a seed does. A seed is a structure with a protective coating that contains more in its inside than a single cell. A seed contains a plant embryo surrounded by a protective coating. An embryo is a multicellular "mini plant" that will grow into a new plant. Along with the embryo is endosperm. Seeds are also much larger than spores.**

24.9 GENERAL CLASSIFICATION SCHEME OF PLANTAE: ATTENTION TO THE MONOCOTS AND DICOTS

- The taxonomic scheme of Plantae is presented in table form.
- Attention is given to the monocots and dicots.
- Monocots form seeds with one cotyledon and dicots form seeds with two cotyledons.

Topic question

What is the difference between a monocot and a dicot seed? **A monocot seed has one cotyledon, or seed leaf, and a dicot has two.**

24.10 PLANT BODY

- Nonvascular plants have root-like, stem-like, and leaf-like structures.
- Vascular plants have vegetative and reproductive organs.

Topic questions

Why do nonvascular plants not have roots, stems, and leaves? **In order for roots, stems, and leaves to be called roots, stems, and leaves, they need to contain vascular tissue. Since the nonvascular plants do not have vascular tissue, they cannot have those structures.**

What are the vegetative structures of a vascular plant? **The roots, stems, and leaves.**

24.11 MERISTEM TISSUE

- Meristem tissue is where plant growth occurs.

Topic questions

Where does plant growth take place? **Meristem tissue.**

What is the difference between lateral meristem and apical meristem tissue? **The only difference is where the meristem tissue is located. Both lateral meristem and apical meristem tissue is where plant growth occurs. Meristem tissue, located specifically at the tips of roots and stems, is apical meristem tissue. Meristem tissue found elsewhere in the plant (such as in the vascular cambium and cork cambium) is called lateral meristem tissue.**

24.12 ROOTS

- Roots anchor a plant to the ground and also absorb nutrients and water from the soil into the plant.
- There are two type of root systems—taproot and fibrous.

Topic question

What is the difference between a fibrous and a taproot system? **A fibrous-root system does not have any one main root but many roots of equal size. A taproot system has one main root with smaller roots growing off the side of the main root.**

24.13 ROOT TISSUES

- There are three layers of a root—the epidermis, cortex, and endodermis.

Topic questions

Where is the vascular cambium located in a root? **It is surrounded by the endoderm and pericycle.**

What is the vascular cambium? **It is where the xylem and phloem are located.**

What is the difference in xylem and phloem location between monocot and dicot roots? **In monocot roots, the xylem and phloem is organized like spokes on a wheel. In dicot roots, the xylem is oriented like an X, and the phloem is in between the arms of the X.**

24.14 STEM

- The stem is divided into nodes and internodes.
- Leaves or branches form at the nodes.
- Stems grow at the apical meristem tissue.

Topic questions

What is significant about the node of a stem? **It is where a leaf or branch grows from a stem.**

What is the difference in location of the xylem and phloem in a monocot and dicot stem? **A monocot stem has the vascular bundles scattered throughout the stem. Dicot stems have the xylem and phloem formed in rings around each other.**

24.15 BARK

- Bark is found only on woody stems. Bark is an added protective layer for the stem.

Topic questions

Why is bark often cracking, splitting, or peeling off a tree? **As the cork cells are made, they soon die and form the outer bark. The way the bark forms is what allows a woody stem to continually grow throughout the life of a woody plant. As the stem grows in circumference, the outer bark cracks. The cork cambium simply fills in the cracks with new cork cells, which die and form new outer bark in the crack. This is the reason bark on trees is cracked and rough.**

What vascular tissue is located in the inner bark layer? **Phloem.**

24.16 GROWTH RINGS

- Tree rings are rings of dead xylem cells.

Topic question

How do tree rings form? **In the springtime, when water is plentiful, a lot of xylem is needed to transport it, so more xylem is made in the spring than in the late summer, fall, and winter. Therefore, the spring xylem is larger and not as densely packed as the later xylem. When a tree is cut down, the spring xylem is lighter in color, and the later xylem is darker in color. During the winter, when no xylem is made, a dark line forms between the late xylem of one season and the spring xylem of the new growing season. The number of rings present indicates how old the tree is, and the thickness of the rings gives an idea of how harsh the climate was for any particular growing season (year).**

24.17 LEAVES

- Leaves are the main photosynthetic organ of the plant.
- Monocot veins run parallel with the leaf. Dicot veins spread out across the leaf in a network from the center vein.
- Leaves have a specific structure that allows them to function optimally during photosynthesis.

Topic questions

What process occurs in the palisade mesophyll of the leaf? **Photosynthesis.**

What are two ways in which a leaf can conserve water (limit transpiration). **The waxy layer on the top of the leaf called the cuticle limits water loss. Also the guard cells on the undersurface of the leaf close when water is not plentiful, so photosynthesis does not occur, and the plant loses less water through transpiration when the guard cells are closed.**

24.18 IMPORTANCE OF WATER FOR PHOTOSYNTHESIS

- Water is critical for photosynthesis to occur.

Topic question

Can photosynthesis occur without water? **No.**

24.19 IMPORTANCE OF WATER FOR TURGOR PRESSURE

- Turgor pressure is the pressure that forms in the plant as a result of water.

Topic question

How does turgor pressure maintain the upright posture of herbaceous stems? **Water is drawn into the large central vacuole by osmosis, causing the vacuole to swell to a large size. This pushes the cytoplasm and organelles up against the plasma membrane making the cell "stiff" or "tight" with the pressure from the large vacuole. When the turgor pressure in all the plant cells is high, the cells all push against one another due to the enlarged vacuoles and maintain the shape of the fleshy parts of the plant.**

24.20 WATER AND NASTIC MOVEMENTS

- Nastic movements are physical movements of the plants due to changes in the plant's environment.

Topic question

What is responsible for nastic movements in plants? **Turgor pressure.**

24.21 WATER AND TRANSPORTATION

- Water is critical for plants to be able to transport nutrients and food throughout the plant.
- Translocation is the process of moving glucose from the leaves to the rest of the plant.

Topic question

Describe how the cohesion-tension theory explains the movement of nutrients throughout the plant. **The cohesion-tension theory starts with water loss in the leaves of the plant. During the daytime, when water is plentiful, the guard cells keep the stoma open so leaves can exchange gas and maximally perform photosynthesis. Photosynthesis causes the plant to lose water. Also, there is evaporation of the water directly from the leaves, called transpiration. Both photosynthesis and transpiration lead to significant water loss in the plant occurring at the leaves, meaning there is an overall deficit of water in the leaves. As water exits the plant from the leaves, the remaining water in the xylem of the branches, trunk, and stem want to stick to the water that is being lost in the leaves. This causes tension to develop because the amount of water in the leaves is less than the amount in the rest of the plant. As the water exits the leaves, the rest of the water in the plant wants to stick to that water; this tension "pulls" the rest of the water molecules up toward the leaves from further down the plant. Since the nutrients absorbed in the roots are dissolved in the water, they go along for the ride, and there is a continual transport of water and nutrients up the plant from the roots.**

24.23 KEY CHAPTER POINTS

- Plants are the primary organisms performing carbon fixation.
- The study of plants is called botany.
- Plant cells are eukaryotic. All plant cells have a cell wall; some have two.
- Plasmodesmata are small tubes that connect plant cells to one another.
- There are two types of plants—vascular and nonvascular.
- Plants can reproduce with seeds and spores, spores only, or neither seeds nor spores.
- There are two types of vascular seed plants—gymnosperms and angiosperms.
- Plants are composed of roots, a stem, and leaves.
- Water is critical for photosynthesis to occur.
- Water is needed for plants to maintain the proper turgor pressure.
- Nastic movements are physical movements of plants that are performed due to changes in their environment.
- Water is critical for the transport of nutrients and glucose in plants.

25 | Plants: Physiology, Reproduction, and Classification

25.0 CHAPTER PREVIEW

In this chapter we will:

- Discuss plant hormones and their effects on plant tissues.
- Discuss asexual means of plant reproduction.
- Study the sexual reproductive cycle of mosses, ferns, conifers, and angiosperms.
- Investigate the structure and function of the flower.
- Discuss the process of pollination and germination.
- Discuss general features of organisms in the Plantae divisions as a wrap-up of our study of plants.

25.1 OVERVIEW

- Plant physiology, reproduction, and discussion of specific phyla are planned for this chapter.

Topic question

25.2 PLANT HORMONES

- Hormones are chemicals used by plants and animals to control certain aspects of the physiology of the organism.

Topic question

What are the five known plant hormones? **They are auxins, gibberellins, cytokinins, abscisic acid, and ethylene.**

25.3 TROPISM

- Tropism is the directional movement of a plant in response to an environmental stimulus.
- Phototropism is the directional growth of a plant toward a light source.
- Thigmotropism is the plant's growth response to touching a solid object.
- Gravitropism is the directional growth of plants directly against gravity.
- Chemotropism is the directional growth of the plant toward a positive chemical stimulus and away from a negative chemical stimulus.

Topic questions

What plant hormone is responsible for most tropisms? **Auxin.**

What is gravitropism? **The growth of a plant directly against gravity.**

How does auxin function to cause the phototropism effect? **Auxin accumulates in the stem opposite the direction of the sun. This causes the cells on the "dark side" of the stem to elongate and bend the top of the stem toward the light.**

25.4 REPRODUCTION: ASEXUAL

- Vegetative reproduction is the term used to describe the formation of a new plant asexually.
- Under the correct conditions, almost any part of a plant—stem, roots, or leaves—is able to form a fully-functional new plant.

Topic questions

What is a scion? **It is a branch or stem cut from a woody plant for the purposes of grafting.**

What is a stolon? **A stolon is a stem that runs along the ground and forms a new plant some distance away from the parent plant. This is one of the ways plants asexually reproduce.**

Which parts of a plant can asexually grow into a new plant? **Any part of a plant can asexually reproduce if the conditions are right.**

25.5 REPRODUCTION: SEXUAL, GENERAL

- Plants demonstrate alternation of generations.
- In nonvascular plants, the gametophyte stage dominates.
- In vascular plants, the sporophyte stage dominates.

Topic questions

When you look at a vascular plant, are you looking at a gametophyte or sporophyte? **A sporophyte. In the vascular plants the sporophyte stage is dominant and the gametophyte is usually so small that it is not easy to see.**

What does it mean when it is said that plants exhibit alternation of generations? **It means that plants have two different forms in which they exist. In one generation, plants live as a sporophyte; in the next generation, plants live as a gametophyte.**

Are gametophytes n or 2n? **They are haploid structures, or n.**

25.6 REPRODUCTION: SEXUAL, MOSSES

- Mosses produce sperm and eggs for sexual reproductive purposes.

Topic question

In which structures do mosses produce the male and female gametes? **The male gametes are formed in the antheridium, and the female gametes are formed in the archegonium.**

25.7 REPRODUCTION: SEXUAL, FERNS

- Fern gametophytes produce sperm in antheridia and eggs in archegonia.
- Fern sporophytes produce spores on the undersurface of fronds.

Topic questions

A fern gametophyte produces what type of reproductive cell? **Gametes—sperm and eggs.**

What are sporangia and what do they do? **Sporangia are spore-producing cells that form on the undersurface of fern fronds. They cluster together to form sori.**

25.8 REPRODUCTION: SEXUAL, GYMNOSPERMS

- The sporophyte is the dominant phase.
- Conifers make male and female cones, which make male and female spores.

Topic question

Describe the process by which a conifer sporophyte produces male and female gametes and forms a diploid seed embryo. **The mature 2n sporophyte produces two separate types of cones on the same tree, male and female. The male cone makes microspores, and the female cone makes megaspores. These spores never leave the cone to mature, but do so while still on the tree. Within the cone, the male spore develops into a haploid male gametophyte, while the female spore develops into a haploid female gametophyte. The male gametophyte is released by the tree and spread by the wind; it will land on the female cone to fertilize the female gametophyte. A diploid zygote is formed and develops into an embryo in the form of a seed. When the seed is mature, the embryo is released from the cone and lands on the ground. If conditions are right, the seed with the embryo will then grow into a diploid sporophyte.**

25.9 REPRODUCTION: SEXUAL, ANGIOSPERMS

- The reproductive structures of angiosperms are contained in flowers.
- Flowers have a complex structure.

Topic questions

What are the female reproductive parts of a flower called? **The pistil.**

What are the parts of the pistil? **The ovary, ovules, style, and stigma.**

What is the male reproductive structure called? **The stamen.**

What are the parts of the stamen? **The filament and the anther.**

What does a megaspore mother cell differentiate into? **An embryo sac, or egg.**

What does a tube cell in pollen do? **When the pollen lands on the stigma, the tube cell forms a tube down the style to the ovary.**

25.10 POLLINATION

- Pollination is the process of transferring pollen from the anther to the stigma.
- A seed contains an embryo and endosperm.
- A fruit is the matured ovary of a plant.

Topic questions

What does the generative cell do when the pollen lands on the stigma? **It divides by mitosis and forms two sperms cells. One of the sperm cells moves down the tube and fertilizes the egg, while the other sperm fertilizes the polar bodies to form the endosperm.**

How do the embryo and endosperm form in a seed? **The zygote differentiates into an embryo, then is encased in a tough protective covering and becomes what is called a seed. The protective covering is called a seed coat. Along with the embryo is a supply of endosperm to serve as nutrition in the early stages of the embryo's growth.**

25.11 GERMINATION

- Once the seed has been dispersed, it needs to have favorable conditions in order to break out of the seed coat and sprout. This is a process called germination.

Topic question

Describe the process of germination. **Since the seed coat is firm and dry, water must soften the seed coat and penetrate into the seed. Once water hydrates the embryo, enzymes are activated that will make the nutrition in the endosperm available to the growing embryo. The seed leaves (cotyledons) emerge from the seed and begin to photosynthesize.**

25.12 SPECIFIC DIVISIONS (PHYLA) OF PLANTAE

- Introduction

Topic question

25.13 NONVASCULAR PLANTS

- Nonvascular plant phyla discussed: Hepatophyta (liverworts), Anthocerotophyta (hornworts), and Bryophyta (mosses).

Topic question

How many species of nonvascular plants are there: 2,000, 17,000, 30,000, or 125,000? **17,000.**

25.14 VASCULAR PLANTS

- Vascular plant phyla are discussed. The seedless vascular species reproduce with spores and include the Divisions of Psilotophyta (whisk ferns), Lycophyta (club mosses), Sphenophyta (horsetails) and Pterophyta (ferns). The seed vascular plants include the Divisions Cycadophyta (cycads), Ginkgophyta (gingkoes), Coniferophyta (conifers), Gnetophyta (gnetophytes), and Anthophyta (angiosperms).

Topic question

How many species of vascular plants are there: 15,000, 75,000, 175,000, or 255,000?
255,000

25.16 KEY CHAPTER POINTS

- Plant hormones are chemicals used by plants to control aspects of their physiology, mainly growth and fruit ripening.
- Auxins, gibberellins, cytokinins, abscisic acid, ethylene, and florigen are all plant hormones.
- All plants are capable of reproducing asexually.
- Plants exhibit alternation of generations in their reproductive cycles.
- Nonvascular plants reproduce sexually with sperm and eggs.
- Ferns reproduce with spores.
- Gymnosperms reproduce with seeds.
- The reproductive structures of angiosperms are contained in flowers.
- Pollination occurs when pollen lands on a stigma.
- Fertilization occurs when the pollen tube grows through the style to the ovary, then the sperm from the pollen fertilizes the egg in the ovary.
- The embryo develops inside the ovary and forms into a seed.
- The ovary matures into a fruit, and the seed is housed inside.
- When a seed begins to grow, it is called germination.
- Nonvascular plant phyla discussed: Hepatophyta (liverworts), Anthocerotophyta (hornworts), and Bryophyta (mosses).
- Vascular plant phyla discussed: the seedless vascular species, which reproduce with spores and include the Divisions of Psilotophyta (whisk ferns), Lycophyta (club mosses), Sphenophyta (horsetails), and Pterophyta (ferns); the seed vascular plants including Divisions Cycadophyta (cycads), Ginkgophyta (gingkoes), Coniferophyta (conifers), Gnetophyta (gnetophytes), and Anthophyta (angiosperms).

26 | Kingdom Animalia I

26.0 CHAPTER PREVIEW

In this chapter we will:

- Introduce the phyla of animals we will be discussing.

- Differentiate vertebrates from invertebrates.

- Define body structure and symmetry terms.

- Discuss the following properties of animals:
    - germ layer/tissue development.
    - embryonic tissue differentiation.
    - segmentation.
    - embryonic development.
    - development patterns.
    - tissue and organ systems.

- Investigate the organisms of the Placozoa and Porifera phyla.

26.1 OVERVIEW

- The organisms of Animalia either have a backbone (vertebrates) or do not have backbones (invertebrates).

- Almost all organisms in Animalia have the following characteristics:
    - a eukaryotic cell structure
    - the ability to reproduce sexually
    - multicellularity with specialization of the cells into tissues that have distinct functions
    - heterotrophic through ingestion and digestion
    - the ability to store carbohydrates in the form of **glycogen** rather than starch
    - no cell walls with special adhesive network outside of the cells called the **extra-cellular matrix**
    - the ability to move through the environment by specialized muscle and nervous tissue; oxygen requiring.

Topic questions

What is a vertebrate? **An animal with a backbone.**

What is an invertebrate? **An animal without a backbone.**

Are more organisms in Animalia vertebrates or invertebrates? **Invertebrates.**

If you are looking at an organism through the microscope and see a cell wall, is it possible this organism is from Animalia? **No. No organisms from Animalia have cell walls.**

26.2 CLASSIFICATION CRITERIA

- Like all other organisms, Animalia is classified based on structures and functions that are similar.

Topic question

26.3 BODY STRUCTURE AND SYMMETRY

- Organisms from Animalia display symmetry.

- Organisms from Animalia display cephalization.

Topic questions

What type of symmetry is displayed when the right half of an organism looks like the left half? **Bilateral symmetry.**

What type of symmetry is shown by aquatic animals, but not land animals? **Radial symmetry.**

What is the head end of a horse called? **The ventral end.**

What part of a horse is the back called? **The dorsal side.**

What is cephalization? **The property that animals display when the head end becomes larger due to the concentration of sensory structures in the head.**

26.4 TISSUES

- Tissues are groups of cells with a similar function.

Topic question

What are the general types of tissues found in animals? **Connective, Epithelium, muscle, and nervous.**

26.5 GERM LAYERS: ENDODERM, MESODERM, AND ECTODERM

- Cells differentiate into tissues during the development period of the animal.
- All animals develop two germ cells (ectoderm and endoderm) and most have a third germ layer (the mesoderm).

Topic questions

What is gestation and when does it begin? **Gestation is the developmental period of an organism, which begins after fertilization.**

How many germ layers do most animals have during their development and what are they called? **Three—endoderm, mesoderm and ectoderm.** What is a zygote? **It is the cell formed from the union of a sperm and an egg.**

26.6 SEGMENTS

- Most animals are segmented.

Topic question

What is segmentation? **The property most animals display in which their bodies are formed by repeated units.**

26.7 COELEM

- Almost all animals have a hollow body cavity, called a coelem.

Topic questions

If an organism shows multicellularity with limited tissues in two layers, it is from which phyla? **Placozoa or Porifera.**

If an animal shows true multicellular tissues in two layers and is radial symmetric, it is from which phyla? **Cnidaria or Ctenophora.**

If an animal is truly multicellular and has tissues in three layers, which phyla could it be from? **All other phyla.**

26.8 EMBRYONIC DEVELOPMENT: GENERAL

- Embryonic development is used to help classify organisms in Animalia.
- Embryology is the study of morphogenesis.

Topic questions

What is morphogenesis? **It is the study of the development of embryos.**

What do embryologists call mitotic cell divisions? **Cleavages.** What is a zygote? **It is the cell formed when a sperm fuses with an egg.**

26.9 EMBRYONIC DEVELOPMENT: BLASTULA

- A blastula is a zygote that has undergone five cleavages and is made up of sixty-four cells.

Topic questions

What is a zygote that has undergone five cleavages and is composed of sixty-four cells called? **A blastula.**

What is a blastocele? **It is the cavity inside the blastula.**

26.10 EMBRYONIC DEVELOPMENT: GASTRULA

- As the blastula grows in size, it begins to fold in on itself. This process is called gastrulation.
- When gastrulation is finished, the embryo is no longer called a blastula but a gastrula.
- As the gastrula grows in size, the germ layers form.

Topic questions

What is the archenteron? **It is the cavity that forms in the gastrula following gastrulation.**

When do the germ layers form? **During the growth of the gastrula.**

What happens to germ layers as the gastrula grows in size? **They differentiate into different types of tissue.**

26.11 EMBRYONIC DEVELOPMENT: NEURAL TUBE

- Vertebrate embryos develop a tube called the neural tube. This becomes the brain and spinal cord.
- Vertebrate embryos also develop a structure just ventral to the neural tube called the notochord. This becomes the spinal column.

Topic questions

What structure does the notochord develop into in vertebrate species? **The spinal column, back bone, or vertebrae.**

What structure does the neural tube develop into in vertebrate species? **The spinal cord and brain.**

26.12 DEVELOPMENT PATTERNS

- Organisms from Animalia show either direct or indirect development patterns once they emerge from the embryonic stage.

Topic questions

What is indirect development? **It is a development pattern characterized by an intermediate stage between the immature and mature organism. In this type of development, the immature organism does not look like the adult organism and goes through an intermediate stage during which time it metamorphs into the adult form of the organism.**

What is direct development? **It is the development pattern in which the immature form of an organism looks like a miniature version of the mature form.**

26.13 VERTEBRATES AND INVERTEBRATES

- There are 950,000 animal species without spinal columns. These are the invertebrate species.
- There are 50,000 animal species with spinal columns. These are the vertebrate species.

Topic questions

What is the main difference between vertebrates and invertebrates? **Vertebrates have a spinal column and invertebrates do not.**

All vertebrate species at some point in their life time have a notochord. True or False? **True.**

26.14 INVERTEBRATES: PHYLUM PLACOZOA

- Placozoa contains only one species, *Trichoplax adhaerens*. It is composed of a couple thousand cells.

Topic question

In which phylum is the most structurally simple animal organism? **Placozoa.**

26.15 INVERTEBRATES: PHYLUM PORIFERA

- Species in Porifera are commonly known as sponges.
- Many people mistake sponges for a plant, but sponges are animals.
- Sponges can sexually and asexually reproduce.
- Sponges are filter feeders.

Topic questions

What is a gemmule? **It is the bud of sponge tissue that forms for asexual reproduction.**

How does a sponge feed? **Sponges are filter feeders. They use flagella to move water and nutrients through openings, called pores, in the body. Sponge cells extract nutrients directly from the water.**

26.17 KEY CHAPTER POINTS

- Although the number of organisms in Animalia are far fewer than the numbers in the kingdoms we have already discussed, "animals" are the most well known.
- The organisms of Animalia are divided into invertebrates and vertebrates.
- The organisms of Animalia are classified based on similarities and differences in their structures and function.
- The organisms in Animalia:
    - are symmetric
    - have tissues
    - are segmented
    - sexually reproduce
    - develop as embryos before emerging as independently functioning organisms
    - store excess energy as glycogen
    - do not have cell walls
    - have cells held together by the extracellular matrix
    - are mobile
    - are heterotrophic
    - are aerobic
- Vertebrates have a "backbone"; invertebrates do not.
- Phylum Placozoa is the most structurally simple phylum in Animalia.
- The phylum Porifera is made up of sponges.

27 | Kingdom Animalia II

27.0 CHAPTER PREVIEW

In this chapter we will:

- Resume our discussion of the Animalia phyla of:
  - Cnidaria
  - Platyhelminthes
  - Nemertea
  - Nematoda
  - Mollusca
  - Annelida
  - Arthropoda
  - Echinodermata
  - Chordata
- Discuss the differences between the vertebrate and invertebrate organisms of Chordata.

27.1 OVERVIEW

- As we proceed through this chapter, we will be discussing progressively more complex organisms, from on organizational and functional standpoint.

Topic question

27.2 INVERTEBRATES: PHYLUM CNIDARIA

- This phylum includes hydras, jellyfish, sea anemones, and corals.
- Organisms of this phylum have two germ cell layers, are capable of sexual and asexual reproduction, have a two-way digestive system and a nerve net. Individual cells perform gas exchange.
- A distinguishing feature of this phylum is the presence of special "stinging" cells called cnidocytes.

Topic questions

What type of symmetry do cnidarians show? **Radial.**

How many germ layers are Cnidarians composed? **Two.**

What is a distinguishing feature of Cnidaria? **The presence of special stinging cells called cnidocytes.**

27.3 INVERTEBRATES: PHYLUM PLATYHELMINTHES

- Organisms from the phylum are also called flatworms.
- They are composed of three germ cell layers, capable of sexual and asexual reproduction, bilaterally symmetric, exhibit cephalization, and have a two-way digestive tract. Individual cells perform gas exchange.
- Several species in the Cestoda and Trematoda classes are important parasites in humans and animals.

Topic questions

How many germ layers do platyhelminths have? **Three.**

Do platyhelminths show cephalization? **Yes.**

How do the platyhelminths perform gas exchange? **Their cells exchange gasses directly with the environment.**

What is a scolex? **It is the specialized head segment of a tapeworm, which is specialized to attach to the host.**

27.4 INVERTEBRATES: PHYLUM NEMERTEA (A.K.A. NEMERTINA)

- Nemertines are commonly called ribbon worms.
- They have all three germ layers, a primitive, but formed, circulatory system, a one-way digestive tract, and a respiratory system.

Topic question

Nemertines are the simplest phylum to exhibit a fully-formed digestive tract and a circulatory system. True or False? **True.**

27.5 INVERTEBRATES: PHYLUM NEMATODA

- Nematodes are also called roundworms and hookworms.

Topic question

What are two genera (plural of genus) of Nematoda, which are important parasites of humans and livestock? *Ascaris* **and** *Enterobius*.

27.6 INVERTEBRATES: PHYLUM MOLLUSCA

- This phylum includes clams, squids, octopi, snails, oysters, and slugs.
- This is a very diverse phylum.
- Most display cephalization, except for the bivalves.
- Most mollusks display indirect development. Also, most mollusks have a true coelem, separate male and female organisms, a one-way digestive tract, and gills to perform gas exchange.
- The Cephalopod class has a well-developed nervous system.

Topic questions

Because Mollusca is a diverse phylum, it is hard to characterize the organisms classified as mollusks. True or False? **True.**

What is a trochophore? **It is a mollusk that is still in the larval stage.**

What is a gill? **It is a specialized tissue that exchanges gas between the organism's blood and its environment.**

27.7 INVERTEBRATES: PHYLUM ANNELIDA

- Annelids are also called earthworms.
- Annelids have a developed nervous system with a small "brain," a one-way digestive system, a closed circulatory system with a small pump, and specialized excretory structures called nephridia. They also reproduce sexually and exchange gasses with the environment across their skin.
- The segmented nature of animals is easy to see in the annelids.

Topic questions

What are nephridia? **Specialized excretory tissue of annelids that remove waste products from the worm.**

What is a crop and gizzard? **They are both part of a one-way digestive system. The crop stores the food, and the gizzard grinds it up to prepare it for absorption.**

Of what system are the aortic arches part? **They are part of the circulatory system.**

27.8 INVERTEBRATES: PHYLUM ARTHROPODA

- This phylum includes insects, spiders, millipedes, centipedes, crabs, lobsters, and horseshoe crabs.
- A distinguishing feature of arthropods is that they have jointed appendages and an exoskeleton made of chitin.
- Most arthropods display indirect development.
- Arthropods have a well-formed nervous system with signals traveling from the body to the brain through a ventral nerve cord, compound eyes to sense light, an open circulatory system, Malpighian tubes to excrete wastes, and variable ways to exchange gas.

Topic questions

What is an exoskeleton? **It is a hard, protective covering on the outside of the organism.**

What is one of the problems with an exoskeleton as it relates to growing? **Since an exoskeleton is rigid, it does not allow for the organism to grow. As an organism grows with an exoskeleton, the exoskeleton peels off at various times to allow for the growth of the organism. This process is called molting. The main key is that the student understands what molting is and it happens as a result of the exoskeleton not allowing growth to occur.**

What is the difference between complete and incomplete metamorphosis? **In complete metamorphosis, the immature form of the organism looks similar to the adult, but is missing some essential features of the adult. In complete metamorphosis, the immature form of the organism looks nothing like the adult form. The immature form goes through several changes, or metamorphoses, before finally assuming the adult form.**

What is a nymph? **It is the intermediate stage of development in an organism that undergoes incomplete metamorphosis.**

What is a pupa? **It is the intermediate stage of an organism that undergoes complete metamorphosis.**

27.9 INVERTEBRATES: PHYLUM ECHINODERMATA

- This phylum includes the sea stars, sand dollars, sea urchins, and sea cucumbers.
- These organisms can reproduce sexually or asexually and exhibit indirect development. They also have three germ layers and a true coelem.

Topic question

What is interesting about the symmetry of the larval and adult forms of the echinoderms? **The larval forms are bilaterally symmetrical, and the adult forms are radially symmetric.**

27.10 INVERTEBRATES: PHYLUM CHORDATA

- The phylum Chordata contains organisms that are both vertebrates and invertebrates.
- This phylum is defined not by the presence of vertebrae, but by the presence of a notochord.

Topic question

What is the defining feature of animals classified in Chordata? **The presence of a notochord at some point in their life cycle.**

27.12 KEY CHAPTER POINTS

- Cnidaria includes hydras, jelly fish, sea anemones, and corals.
- The organisms of Platyhelminthes are called the flatworms.
- The organisms of Nemertea are called the ribbon worms.
- The organisms of Nematoda are called the roundworms.
- Mollusca includes squids, octopi, snails, oysters, slugs, and clams.
- Annelida includes earthworms and leeches.

- Arthropoda includes insects, trilobites, centipedes, millipedes, spiders, scorpions, crabs, lobsters, shrimp, and barnacles.

- Echinodermata includes sea stars, sand dollars, sea urchins, and sea cucumbers.

- Chordata includes tunicates, lancelets, sharks, fish, amphibians, reptiles, birds, mammals, and humans.

- The presence of a dorsal nerve tube and a notochord separate Chordata from all other phyla in Animalia.

28 | Kingdom Animalia III

28.0 CHAPTER PREVIEW

In this chapter we will:

- Discuss the features of vertebrates.

- Discuss vertebrate organ systems.

- Review the features of the organisms classified into:
  - Agnatha
  - Chondrichthyes
  - Osteichthyes
  - Amphibia
  - Reptilia

- Introduce the concept of the one-loop and two-loop circulatory systems.

- Discuss the function of capillaries.

- Investigate the structure and function of the amniotic egg.

- Discuss endothermia and ectothermia.

28.1 OVERVIEW

- The defining feature of all vertebrates is the presence of a dorsal spinal cord protected by a vertebral column.

Topic question

What is the defining feature of organisms classified in Vertebrata? **The presence of a dorsal spinal cord protected or surrounded by a spinal column.**

28.2 EVOLUTION OF LAND ANIMALS?

- Evolutionists believe that land animals evolved from aquatic animals.

- Creationists believe that all animals were created to occupy their current environment.

Topic question

Why is it that evolutionists continue to promote evolution based on faulty genetic understanding? **Because evolution is a way to explain the origin of life without God's involvement. Since most leaders of evolutionary thought are, or were, atheists or agnostics, it is impossible to believe that God is the reason all life is here when you specifically do not believe God exists.**

28.3 ORGAN SYSTEMS

- All vertebrates have highly-ordered organ systems.

- The organ systems vertebrates have are reproductive, neurological, digestive, circulatory, respiratory, excretory (exocrine), endocrine, musculoskeletal, immune, and connective.

Topic question

What are the vertebrate organ systems discussed in this course? **Reproductive, neurological, digestive, circulatory, respiratory, excretory, endocrine, musculoskeletal, immune, and connective.**

28.4 PHYLUM CHORDATA, CLASS AGNATHA

- Agnatha is also known as the jawless fish.
- They have a well-developed lateral line system

Topic question

What are the two defining characteristic of the organisms of Agnatha? **They do not have a jaw and the notochord remains throughout the organism's life.**

28.5 PHYLUM CHORDATA, CLASS CHONDRICHTHYES

- Organisms classified in Chondrichthyes are the sharks, rays, and skates. They have skeletons made out of cartilage.
- All organisms in this Class are carnivores, exhibit cephalization.

Topic questions

What is a defining feature of Chondrichthyes, as in sharks? **Their skeleton is made out of cartilage.**

How do Chondrichthyes exchange gasses? **Gills.**

28.6 PHYLUM CHORDATA, CLASS OSTEICHTHYES

- These organisms are the boney fish.
- Some fish species demonstrate parental care.

Topic questions

What type of circulatory system do fish have? **Closed.**

How many chambers does their heart have? **Two.**

Describe the circulatory pathway in a fish, starting in the atrium. Is it a one-loop or two-loop system? **The circulatory system of a fish is a one-loop system. Starting in the atrium, the blood flows from the atrium to the ventricle, to the gills, to the tissues, then from tissues back to the atrium.**

28.7 PHYLUM CHORDATA, CLASS AMPHIBIA

- Amphibians are land-living animals. This class includes frogs, toads, salamanders, and caecilians.
- They have a close reliance on water, because some amphibians need wet skin to exchange gasses, and all amphibians rely on water for their reproductive cycles.
- Because of the presence of lungs, the circulatory system is now a two-loop system.

Topic questions

What are the two-loops called of a two-loop circulatory system? **One-loop is the pulmonary loop or pulmonary circulation and the other is the systemic loop or systemic circulation.**

Describe the path that blood takes in the frog , starting from the right atrium. **The blood is pumped from the right atrium into the single ventricle, to the lungs, to the left atrium, to the ventricle, to the tissues, then back to the right atrium.**

28.8 PHYLUM CHORDATA, CLASS REPTILIA

- Reptiles are diverse. This Class includes snakes, lizards, alligators, crocodiles, and turtles.
- Unlike most of the species of vertebrates discussed previously, reptiles have internal fertilization, and a tough shell forms around the embryo.
- The type of egg formed by reptiles is called the amniotic egg.

Topic questions

Why are reptiles less dependent on water than amphibians? **They have watertight skin, so they do not need to keep their skin wet all the time like amphibians. Also, they produce** amniotic eggs, which protect the developing embryo from drying out. Amphibians must lay their unprotected eggs in or close to water.

What is the structure of the amniotic egg? **The shell is hard, but porous—to allow gasses to pass through. The amnion surrounds the embryo. The yolk sac is attached to the embryo and provides nutrition for the growing embryo. The allantois stores nitrogen wastes. Albumin serves as a source of protein.**

28.9 PARITY

- In oviparity, a shell forms around an embryo, which is internally fertilized and then laid.

- In ovoviviparity, a shell forms around an embryo that has been internally fertilized, but the egg is retained for a while. The egg is laid before the embryo hatches, or is retained and the embryo hatches while still in the mother.

- In viviparity, no shell forms around the embryo. There is internal fertilization, and the embryo is retained in the mother, receiving nutrition and performing gas exchange directly from the mother's blood stream through the placenta.

Topic question

What is the difference between oviparity, ovoviviparity, and viviparity? **In oviparity, a shell forms around an embryo, which is internally fertilized and then laid. In ovoviviparity, a shell forms around an embryo that has been internally fertilized, but the egg is retained for a while. The egg is laid before the embryo hatches or is retained, and the embryo hatches while still in the mother. In viviparity, no shell forms around the embryo. There is internal fertilization, and the embryo is retained in the mother, receiving nutrition and performing gas exchange directly from the mother's blood stream through the placenta.**

28.10 ECTOTHERMS AND ENDOTHERMS

- Ectotherms do not have a very high metabolism rate and are unable to generate their own body heat. They need to rely on their environment in order to maintain homeostatic temperature.

- Endotherms have a high metabolic rate. Endotherms do not depend on the environment to maintain a homeostatic temperature.

Topic question

What is the difference between an endotherm and an ectotherm? **Endotherms generate enough heat through their metabolic processes to maintain temperature homeostasis regardless of the environmental temperature. Ectotherms do not have as high a metabolic rate and must rely on the environment to regulate their body temperature.**

28.12 KEY CHAPTER POINTS

- A vertebrate is an animal with a dorsal spinal cord protected by a spinal column.

- Evolutionists believe that land-based animals evolved from aquatic animals. Creationists believe all animals were created to occupy whatever environment they currently occupy.

- Vertebrates have tissues that are highly developed into organ systems.

- Agnatha includes hagfishes and lampreys.

- Chordata includes many classes with which we are all familiar—Chondrichthyes includes the sharks, rays, and skates; Osteichthyes includes the boney fish; Amphibia includes the amphibians; and Reptilia includes the reptiles.

- There are three different types of embryo development among vertebrates—oviparity, ovoviviparity, and viviparity.

- Ectotherms have a low metabolic rate and do not produce body heat. They are dependent on their environment to maintain a homeostatic temperature.

- Endotherms have a high metabolic rate and produce their own body heat. They are not dependent on their environment to maintain a homeostatic temperature.

29 | Kingdom Animalia IV

29.0 CHAPTER PREVIEW

In this chapter we will:

- Discuss the classification of birds into the phylum Aves.
- Review the evolutionary paradigm that birds evolved from reptiles and the evidence against that occurring.
- Investigate the bird structure and function of feathers and air sacs.
- Discuss the classification of mammals into three classes—Monotremata, Marsupialia, and Placentalia.
- Discuss common mammalian traits.
- Investigate mammalian specializations of:
  - flight
  - aquatic life
  - echolocation
  - brain enlargement
- Introduce human organ systems.

29.1 OVERVIEW

- Organisms classified in Mammalia and Aves are all endotherms.

Topic question

Some of the organisms classified in Aves are not endotherms. True or False? **False. All birds are endotherms.**

29.2 PHYLUM CHORDATA: CLASS AVES

- Ornithology is the study of birds.

Topic question

What features do all birds have in common? **Endothermia, feathers, wings, beaks, oviparity in an amniotic egg, specialized respiratory system, and lightweight skeleton.**

29.3 BIRD EVOLUTION? ARCHAEOPTERYX AND BIRDS

- *Archaeopteryx* is a bird fossil, not an intermediate form between birds and dinosaurs.

Topic question

What is the general consensus regarding *Archaeopteryx*? **It is a bird fossil. Almost all leading ornithologists who have studies it think it is a bird.**

29.4 BIRD EVOLUTION? FEATHERS

- Evolutionists believe that feathers are modified scales.
- Scales are small outpouchings of skin. Feathers are exceedingly complex structures that develop from follicles.
- Scales are made from a-keratins; feathers are made from f-keratins.

Topic question

Why is it hard to believe that feathers are modified scales? **If feathers are modified scales, then they should have something in common with scales. Scales are simple folds of skin. Feathers are an unbelievably complex structure from a structural standpoint. Scales and feathers are not even made from the same protein and scales are essentially skin while feathers grow out of a specialized structure of the skin called a follicle.**

29.5 AVES: CHARACTERISTICS

- All members of Aves have the following characteristics in common:
  - wings
  - amniotic egg with oviparity
  - well-formed cerebellum to coordinate flight
  - one-way digestive system with common opening to eliminate waste
  - four-chambered heart with two-loop system
  - specialized lightweight body plan
  - excrete wastes as urea.

Topic question

Describe the special way birds breathe. **Air enters into the bird through the nostrils and goes not only into the lungs, but also fills the air sacs. During inspiration (breathing in), the lungs extract oxygen and release carbon dioxide while the air sacs are filled with air (no gas exchange occurs in the air sacs as they simply store the air during inspiration). During exhalation (breathing out), the air in the lungs, now containing carbon dioxide, rushes out of the lungs and is replaced with the air from the air sacs, which still has oxygen in it. This allows the lungs of birds to continually exchange gas as the lungs are continually filled with oxygen-containing air. In addition, the air sacs help to significantly reduce the weight of the bird.**

29.6 PHYLUM CHORDATA: CLASS MAMMALIA

- Mammals are divided into:
  - Monotremata, "the egg laying mammals"
  - Marsupialia, "the pouched mammals"
  - Placentalia, "the placental mammals" (consists of over 95% of the mammal species)

Topic question

What are the three general categories of Mammalia? **Monotremata, "the egg laying mammals"; Marsupialia, "the pouched mammals"; Placentalia, "the placental mammals" mammals.**

29.7 MAMMALIA: CHARACTERISTICS

- Mammals have the following characteristics:
  - mammary glands
  - hair
  - the ability to store extra energy as fat
  - a single, lower jaw bone
  - specialized teeth
  - a high degree of parental care
  - a large brain-to-body size ratio
  - heterotrophic
  - a four-chambered heart with a two-loop circulatory system
  - lungs

Topic questions

If you were looking at a skull that had only one jaw bone, could it be from a mammal? **Yes, mammals all have a single jaw bone.**

Do mammals that live in the water, like whales, have gills? **No, all mammals have lungs, including the ones that are aquatic.**

29.8 MAMMALIA: SUB-CLASS MONOTREMATA

- Monotremes lay eggs with soft, leathery shells.

Topic question

There are many species of monotremes that are not extinct. True or False? **False. There are only two species—the duck-billed platypus and the anteater.**

29.9 MAMMALIA: SUB-CLASS MARSUPIALIA

- Almost all marsupials live in Australia.

- They have pouches to protect their young, which are born at an immature age and develop in the pouch.

Topic question

A mammal that gives birth to an immature baby and raises it in a pouch is classified in which sub-class? **Marsupialia.**

29.10 MAMMALIA: SUB-CLASS PLACENTALIA

- The defining characteristic of placental mammals is the way in which their embryos develop. The embryos are retained inside of the female. The fetus receives its nourishment and exchanges gas and wastes with the mother's bloodstream through the placenta.

Topic question

How and when does the placenta form? **It forms just after fertilization, when the embryo attaches onto the lining of the uterus. The placenta then grows from the embryo and the uterus.**

29.11 PLACENTALIA: SPECIALIZATIONS, AIR

- Bats, from the Order Chiroptera, are the only mammals specialized to fly.

Topic question

What mammalian order is the only one which has the ability to fly? **Chiroptera.**

29.12 SPECIALIZATIONS: WATER

- Some mammals rely on water for survival, but are not obligated to remain in the water.

- Organisms from the Orders Sirenia and Cetacea are exclusively aquatic because they cannot survive out of the water.

- Water adaptations include: blow holes; flippers; flukes; large-capacity lungs; and tissues specialized to extract the maximum amount of oxygen possible.

Topic questions

What are the specialized structures, or physiologic processes, that allow organisms from Sirenia and Cetacea to live in the water? **Blow holes; flippers; flukes; large-capacity lungs; tissues specialized to extract the maximum amount of oxygen possible.**

Since they live in the water, aquatic mammals do not have hair, nor do they feed their young from mammary glands. True or False? **False. The aquatic mammals, as all placental mammals, feed their young with mammary glands after they are born and form a placenta.**

29.13 SPECIALIZATIONS: ECHOLOCATION

- Echolocation is the use of sound waves to detect or locate objects in their environments.

Topic question

What mammalian orders use echolocation? **Cetacea and Chiroptera.**

29.14 SPECIALIZATIONS: BRAIN

- Mammals have large brain-to-body ratios. The portion of the brain that is enlarged the most is the cerebrum, or the thinking part of the brain.

Topic question

What part of a mammal's brain has enlarged the most to account for the large brain-to-body size ratio? **The cerebrum**.

29.15 INTRODUCTION TO HUMAN SYSTEMS

- The structure of the human system is called anatomy.

- The function of the human system is called physiology.

- The systems we will study in human anatomy and physiology are excretory, neurologic, endocrine, immune, circulatory, respiratory, digestive, musculoskeletal, and connective.

Topic question

29.17 KEY CHAPTER POINTS

- This chapter focuses on the classes Aves (birds) and Mammalia (mammals).

- The study of birds is called ornithology.

- Archaeopteryx is not a transitional form between birds and dinosaurs. It is a bird.

- Feathers could not possibly have evolved from scales because there is nothing similar about scales and feathers.

- Birds have: amniotic eggs; high parental behavior; lightweight skeleton designs; wings; a large cerebellum to coordinate flight. They are also all endothermic.

- Mammals are divided into egg-laying mammals (Order Monotremata), pouched mammals (Order Marsupialia), and placental mammals (Order Placentalia).

- Mammals have: mammary glands; hair; extra energy stored as fat; a single, lower jaw bone; specialized teeth; a high degree of parental care.

- Many mammal species have specialized adaptations: wings, fins, blow holes, echolocation, and large brains.

30 | Human Anatomy and Physiology I
Nervous System and Special Senses

30.0 CHAPTER PREVIEW

In this chapter we will:

- Describe the anatomy and physiology of the nervous system, with attention to the:
  - individual nerve cell (neuron)
  - neuron's ability to generate and conduct an electrical impulse
  - neurotransmitters
  - central nervous system
  - peripheral nervous system
  - autonomic nervous system

- Describe the anatomy and physiology of vision, hearing, taste, and smell.

30.1 OVERVIEW

- Humans are unique because of our ability to think and reason. However, our basic anatomy and physiology are similar to other mammalian species.

Topic questions

What is anatomy? **It is the structure of living things.**

What is physiology? **It is how the anatomy functions.**

30.2 NERVOUS SYSTEM: GENERAL

- The nervous system is a complicated network of cells that function to sense changes in the environment and transmit the information to the brain. Once the information is received by the brain, the nervous system coordinates how the organism responds to the information.

Topic question

What is the basic purpose of the nervous system? **It receives information from the environment, then processes it to react in some way to the information presented to it.**

30.3 NERVOUS SYSTEM: NERVE CELLS AND TISSUES

- Nerve tissue is composed of nerve cells and supporting tissue called glial cells.

- Nerve cells send and receive electrical impulses.

- Glial cells support and protect the nerves.

- There are three basic types of neurons—sensory neurons, motor neurons, and interneurons.

- The brain and spinal cord are called the central nervous system, and the sensory and motor nerves are called the peripheral nervous system.

- The thing or event that causes a nerve to generate an impulse is called a stimulus.

Topic questions

What is a glial cell? **It is a support cell of the nervous system.**

What is the general structure and function of each part of a neuron? **The neuron is composed of a cell body containing the nucleus. This is where the proteins are made. Each neuron has one axon extending from it, which carries electrical signals away from the cell body. The axon is covered by fat (myelin), which makes the axon conduct the impulse faster. There are many dendrites that are smaller projections off the cell body. These carry electrical impulses to the cell body.**

What parts make up the central nervous system (CNS)? The peripheral nervous system (PNS)? **The CNS is the brain and the spinal cord. The PNS is the sensory and motor nerves.**

30.4 NERVE PHYSIOLOGY: GENERAL

- At rest, the nerve is polarized. When it is stimulated to generate an impulse, it depolarizes.
- The resting potential is maintained through the action of the Na-K-ATP pump.

Topic questions

What is the Na-K-ATP pump and how does it maintain the resting potential? **The Na-K-ATP pump actively pumps sodium out of the neuron and potassium into the neuron. This maintains a resting potential, or charge, of -70mV on the inside of the nerve cell as compared to the outside of the cell.**

How does the cell establish the resting potential by using the pump? **The cell pumps out many positively-charged sodium ions and pumps in fewer positively-charged potassium ions. This results in more positive ions accumulating outside of the neuron membrane. The inside of the neuron is maintained as more negative than the outside of the neuron.**

30.5 NERVE PHYSIOLOGY: DEPOLARIZATION AND REPOLARIZATION

- When a nerve receives the proper stimulus, it depolarizes.
- When a nerve depolarizes, it allows sodium to rush into the neuron and potassium to rush out of the neuron.
- When the nerve is finished conducting the impulse, it repolarizes.

Topic questions

How does a nerve depolarize? **The electrical charge of the neuron changes. The membrane becomes permeable to sodium and impermeable to potassium. Sodium rushes into the neuron, and the membrane potential changes to positive in the area where the sodium has entered the neuron.**

What is an action potential? **It is a wave of depolarization along a neuron.**

What is repolarization? **When the nerve is finished conducting the action potential, the Na-K-ATP pump quickly pumps sodium out of the neuron and potassium into the neuron. This re-establishes the resting potential of the neuron.**

30.6 NERVE PHYSIOLOGY: ALL OR NONE

- When a neuron is stimulated, it will either conduct an impulse or it won't. This means the neuron will either depolarize and generate an action potential or it won't. This is the all-or-none response.

Topic questions

What happens if a stimulus does not reach threshold for a neuron? **Nothing. If the stimulus does not reach threshold, the nerve will not depolarize.**

What is the all-or-none response? **Either a stimulus reaches threshold or it does not. There is no in-between stimulus. The neuron either fires or it does not.**

30.7 NERVE PHYSIOLOGY: SALTATORY CONDUCTION

- Electrical impulses "jump" from one node of Ranvier to another down the length of the neuron. This is called saltatory conduction

Topic question

What is saltatory conduction and what makes it possible. **It is the jumping of an electrical impulse from node of Ranvier to node of Ranvier. This is possible because of the insulating effects of myelin.**

30.8 NERVE PHYSIOLOGY: BRUSH

- The brush is the area of the neuron where the axon ends in a branched network to interact and communicate with dendrites of other nerves.

- Neurotransmitters are chemicals nerves use to communicate with one another.

Topic question

How do nerves communicate with one another? **When the impulse travels down the axon and nears the brush, the calcium gates open and allow the calcium ions (which are in much greater concentration outside the cell than inside) to rush into the axon in the area of the brush. Once the calcium is inside of the axon, it stimulates the vesicles containing the neurotransmitters to move toward the membrane. The vesicles fuse with the membrane, then the neurotransmitters inside are dumped into the synaptic cleft though the process of exocytosis. The neurotransmitters diffuse across the synaptic cleft and stimulate the dendrites.**

30.9 NERVE PHYSIOLOGY: EXCITATORY AND INHIBITORY TRANSMITTERS

- Excitatory neurotransmitters are so named, because they increase the conductance of the neuron. This means it is easier for the nerve to conduct an impulse.

- Inhibitory neurotransmitters are so named because they decrease the conductance of the interneuron system. This means they make it harder for the nerve to conduct an impulse.

Topic question

Why is an inhibitory neurotransmitter inhibitory? **Because an inhibitory neurotransmitter makes it harder for the nerve to conduct an impulse.**

30.10 NERVE SYSTEM ANATOMY: PNS AND CNS

- The CNS consists of the brain and spinal cord.

- The PNS consists of the peripheral nerves.

Topic question

30.11 NERVE SYSTEM ANATOMY: PROTECTIVE TISSUES

- There are three protective tissues around the brain. The tissues are called meninges.

Topic question

What are the layers of the meninges? **The dura mater, the arachnoid, and the pia mater.**

30.12 NERVE SYSTEM ANATOMY: CNS

- The brain is surrounded by fluid called CSF and is divided into different parts that are called lobes.

- The general areas of the brain are the forebrain, midbrain, and hindbrain.

- Grey matter is where the cell bodies of brain neurons are located.

- White matter is where the myelinated axons are located.

Topic questions

What is the forebrain also called? **The cerebrum.**

What parts of the brain make up the hindbrain? **The brain stem and the cerebellum.**

Which part of the brain controls and fine-tunes movement? **The cerebellum.**

Which part of the brain is involved with thinking and motor control? **The frontal lobes.**

What is contained in gray matter of the brain? **The cell bodies of brain axons.**

30.13 NERVE SYSTEM ANATOMY: PNS

- The PNS is composed of spinal nerves and peripheral nerves.

Topic questions

What do motor nerves innervate and what do they do? **Motor nerves innervate muscle and make you move.**

What do sensory nerves innervate and what do they do? **Sensory nerves innervate the skin and make you feel things.**

30.14 NERVE SYSTEM ANATOMY: AUTONOMIC NERVOUS SYSTEM

- The ANS controls involuntary functions such as the heartbeat, digestion, sweating, etc.
- There are two components of the ANS, the sympathetic and parasympathetic nervous systems.

Topic questions

Which part of the ANS gets you ready for fight or flight? **The sympathetic nervous system.**

30.15 SPECIAL SENSES: VISION

- Vision is possible because of eyes.
- Eyes have a perfect structure for seeing.

Topic questions

Which cells allow you to sense light? **Rods.**

Which cells allow you to sense color? **Cones.**

In the correct order, what structures of the eyes does a ray of light pass through on the way to the brain? **The cornea, aqueous humor, pupil, lens, vitreous, retina, optic nerve and, finally, the brain.**

30.16 SPECIAL SENSES: HEARING

- Hearing is possible through the complex interactions of air, a membrane, bones, fluid, and a nerve.

Topic question

Describe the mechanism of hearing. **Sound waves pass into the auditory canal and strike the ear drum (tympanum). The ear drum is connected to the three ear bones (ossicles); when the sound hits the ear drum it vibrates. When the ear drum vibrates, it causes the ossicles to vibrate. The ossicles transmit the vibration into the cochlea. The cochlea is filled with fluid. When the ossicles vibrate, the fluid in the cochlea moves. The movement of the fluid causes movement of tiny hairs on the ends of nerve cells. As the hairs move, nerve impulses are generated, then transmitted through the acoustic nerve to the brain and processed.**

30.17 SPECIAL SENSES: TASTE AND SMELL

- The sense of taste is called the gustatory sense.
- The sense of smell is called the olfactory sense.

Topic questions

How many sensations can the tongue sense? **Four and possibly five.**

What are they? **Sweet, salty, sour, and bitter. The fifth possible one is umami.**

30.19 KEY CHAPTER POINTS

- Humans are unique mainly because of our ability to think and reason.
- However, much of the anatomy and physiology of humans is similar to all other mammals.

- The nervous system components are the brain, spinal cord, peripheral nerves, eyes, ears, nose, and tongue.
- Nerve tissue is made up of nerve cells and support cells (called Glial cells).
- The nervous system is divided into the central nervous system (CNS), the peripheral nervous system (PNS), and the autonomic nervous system (ANS).
- When a nerve is at rest, it is polarized. When it conducts an impulse, it depolarizes in an all-or-none fashion.
- Neurotransmitters are chemicals that nerves use to communicate signals to one another.
- The parts of the nervous system have specific names and functions.
- The CNS includes the brain and the spinal cord.
- The PNS includes the motor and sensory nerves.
- The ANS includes the nerves that perform our automatic functions.
- The special senses are sight, hearing, taste, and smell.

31 Human Anatomy and Physiology II
Immunologic, Endocrine, and Excretory Systems

31.0 CHAPTER PREVIEW

In this chapter we will:

- Describe the components and function of the immune system.
- Discuss the specific and nonspecific phases of the immune response.
- Define active and passive immunity.
- Discuss the condition of an autoimmune disease.
- Describe the anatomy of the endocrine system.
- Discuss the effects and control of the endocrine system.
- Investigate the anatomy and physiology of the excretory system.

31.1 OVERVIEW: IMMUNE SYSTEM

- The immune system is the system the body has in place to protect against invaders.

Topic questions

What is a pathogen? **An infectious organism; something that can cause disease.**

What are some common pathogens the immune system fights off? **Viruses, bacteria, parasites, and fungi.**

31.2 IMMUNE SYSTEM: BARRIERS

- Physical and chemical barriers exist as the first line of defense in the immune system.

Topic question

What are some specific examples of chemical barriers to prevent pathogens from entering the body? **They are found mainly in the fluids of the eyes, mouth, and stomach. These chemicals are usually enzymes that function to break down the cell wall or membrane of the pathogen. They are present all the time, on standby in case a pathogen gets past the physical barriers. In addition, the acidic environment of the stomach often kills pathogens.**

31.3 IMMUNE SYSTEM: NONSPECIFIC RESPONSE

- The nonspecific response consists of an inflammatory reaction, fever, and the activation of immune cells called phagocytes.

Topic questions

What is histamine and what does it accomplish in the nonspecific response? **Histamine is a chemical that is released when tissues are injured. It results in increased blood flow to an injured area because it causes blood vessel dilation. Histamine also results in phagocytes being activated.**

What are phagocytes? **They are cells that remove foreign material by engulfing it.**

31.4 IMMUNE SYSTEM: SPECIFIC RESPONSE; IMMUNE RESPONSE

- The specific response to an invader is also called the immune response.
- The immune response is carried out by special white blood cells called lymphocytes.
- The reason this is called the specific response is because a specific pathogen is targeted by the immune system for removal.

Topic questions

What is an antigen? **It is usually a protein that causes the specific response of the immune system to be activated.**

Describe how an immune response is activated. **A foreign invader escapes the barriers and the nonspecific response. Macrophages engulf and process the invader and present antigens from the invader to the t-helper cell. The t-helper cell takes the antigen to the b-cells and cytotoxic t-cells. This allows the b and t-cells to identify the invader and prepare to specifically attack it. The b-cell makes antibodies that will be secreted into the blood and stick to the antigen on the pathogen. The cytotoxic t-cell will make special molecules on its cell surface called receptors, which will stick to the antigen on the pathogen. Many b-cells and cytotoxic t-cells are cloned, then all attack only the pathogen with the antigen on it. The antibodies make it easy for neutrophils and macrophages to target and engulf the pathogens. The cytotoxic t-cells stick to the pathogens and kill them. In this way, the pathogen is specifically targeted for removal.**

What is memory? **The immune system retains the ability to quickly respond to a pathogen that it has already mounted an immune response to in the past. If the body is exposed to the same pathogen in the future, it can quickly remove it due to the memory response.**

31.5 IMMUNE SYSTEM: ACTIVE AND PASSIVE IMMUNITY

- Immunity which someone develops as the result of being exposed (infected) by a pathogen is called active immunity.
- Passive immunity is immunity that exists due to the transfer of antibodies from one person to another.

Topic question

What type of immunity is acquired as a result of immunizations? **Active.**

31.6 IMMUNE SYSTEM: AUTOIMMUNE RESPONSE

- Sometimes people mount an immune response against their own bodies. This is known as an autoimmune response or disease.

Topic question

What tissue is targeted in the autoimmune disease of childhood diabetes mellitus (type I diabetes). **The pancreas.**

31.7 ENDOCRINE SYSTEM: GENERAL

- The endocrine system is a complex system, which uses chemical messengers, called **hormones**, so one organ can communicate with another organ some distance away. Hormones are released into the blood stream by the organ that has synthesized it, referred to as a **gland**, then travels to the organ it is meant to communicate with, called the **target organ**.

Topic questions

What is a hormone? **Hormones are the chemical messengers of the endocrine system.**

What is a target organ and what do they have for hormones? **It is the organ with which the hormone is meant to communicate. Every target organ has a receptor on it that sticks to the hormone. When the hormone passes by the target organ, it binds to the receptor.**

31.8 ENDOCRINE SYSTEM: HORMONES

- There are two types of hormones—amino acid hormones and steroid hormones.

Topic question

What is the difference between amino acid and steroid hormones? **Amino acid hormones are not soluble in the cell membrane. The receptors for the amino acid hormones** are on the cell surface of the target organ cells. Steroid hormones are soluble in the cell membrane. They are able to pass through the membrane and enter the cell. The receptors for steroid hormones are in the cytoplasm of the target organ cells.

31.9 ENDOCRINE SYSTEM: REGULATION

- The hypothalamus is the super-controller of the endocrine system.
- The pituitary gland secretes many different types of hormones into the blood as directed by the hypothalamus.
- The endocrine system is regulated by feedback inhibition.

Topic question

Describe how feedback inhibition works using an example from the chapter. **The hypothalamus detects the level of thyroid hormones in the blood stream. If the level of thyroid hormones drops too low, the hypothalamus senses it and releases thyrotropin-releasing hormone (TRH). TRH causes the pituitary to release the thyroid stimulating hormone (TSH) into the blood. TSH circulates in the blood, then binds to the TSH receptors in the thyroid gland. Once these receptors are bound by TSH, the thyroid gland is stimulated to release its thyroid hormones into the blood. Eventually, the levels of the thyroid hormones reach a certain "normal" level in the body. This normal level gives feedback to the hypothalamus that there are enough thyroid hormones in the body. This causes the hypothalamus to stop releasing TRH into the blood which, in turn, causes the pituitary not to release TSH into the blood. That causes the thyroid to not release any more of the thyroid hormones. When the level of the thyroid hormones gets low enough again, the hypothalamus receives feedback that the body needs more thyroid hormones, and the whole cycle starts over.**

31.10 EXCRETORY SYSTEM

- The kidney filters blood to remove wastes and conserve water, ions, and nutrients.
- The functional unit of the kidney is the nephron.

Topic question

Describe how the process of blood filtration occurs in the kidney. **Blood is carried to the kidney through the renal artery. Once inside the kidney, the renal artery flows into many glomeruli. The cells of the glomerulus allow protein waste products (mainly urea), ions (like sodium, calcium, and potassium), water, and nutrients to filter out of the blood in the glomerulus and into Bowman's capsule. This filtered material is called filtrate. From Bowman's capsule, the filtrate moves into the collecting duct on its way to the ureters and the bladder. As the blood continues flowing through the glomerulus, it passes by the collecting duct and re-absorbs nutrients, water, and ions from the filtrate. In doing so, the substances the rest of the body needs are taken up into the blood again. Nothing that is needed is lost. Urea and other waste products remain in the collecting duct and continue to pass from the urine-collecting duct system into the ureters and the bladder. Once the bladder is full, you go the bathroom and empty the urine form the bladder.**

31.12 KEY CHAPTER POINTS

- All organisms have ways to fight off foreign invaders. In humans and other animals, it is called the immune system.
- There are many barriers—physical, chemical, nonspecific, and specific—that are present to prevent entry of pathogens into the body.
- The specific response is the immune response.
- The immune response results in active immunity.
- The endocrine system is a complex system that uses chemical messengers to communicate with organs and tissues far from one another.
- The excretory system is the way the body eliminates waste materials.

32 Human Anatomy and Physiology III
Circulatory and Respiratory Systems

32.0 CHAPTER PREVIEW

In this chapter we will:

- Discuss the components of the circulatory system.
- Review the two-loop circulatory system.
- Examine the internal and external heart anatomy.
- Investigate the electrical activity of the heart and how it relates to EKG activity and the generation of blood pressure.
- Study the structure and function of blood (even though it is part of the connective tissue system, many students find it makes more sense to discuss it as part of the circulatory system).
- Discuss the anatomy of the respiratory system.
- Investigate how inhalation and exhalation occur.

32.1 CIRCULATORY SYSTEM: GENERAL

- The purpose of the circulatory system is to move oxygen, hormones, and nutrients to tissues/organs and carbon dioxide and wastes from tissues/organs.
- Humans and all vertebrates have a closed, two-loop circulatory system with a four-chambered heart.

Topic question

What are the components and functions of the circulatory system? **The heart pumps the blood through the blood vessels. Arteries carry blood away from the heart. Veins carry blood toward the heart. Capillaries are small vessels where gas and nutrient exchange occur. Blood is technically not part of the circulatory system, but its contribution to the circulatory system cannot be ignored.**

32.2 CIRCULATORY SYSTEM: ANATOMY

- There are two-loops to the system—the peripheral (systemic) and the pulmonary loops.
- The heart has four chambers and four valves.
- The pulmonary arteries carry blood from the heart to the lungs.
- The pulmonary veins carry blood from the lungs to the heart.

Topic questions

What are the two large and muscular chambers of the heart called? **Ventricles.**

What are the two small, receiving chambers of the heart called? **Atria.**

What is the flow of blood through all major structures that we have learned the names of starting from the right atrium? Be sure to include the heart valves. **Right atrium, tricuspid valve, right ventricle, pulmonic valve, pulmonary arteries, lungs, pulmonary veins, left atrium, mitral valve, left ventricle, aortic valve, aorta, systemic circulation, capillary beds, veins, inferior, or superior vena cava, right atrium.**

Is the blood in the pulmonary vein oxygenated or deoxygenated? **It is oxygenated because it has gone through the lungs and released carbon dioxide and picked up oxygen.**

32.3 CIRCULATORY SYSTEM: PHYSIOLOGY OF A HEARTBEAT

- The electrical signal for the heartbeat is generated in the S-A node.
- The A-V node transmits the heartbeat from the atria into the ventricles.
- Systole is when the ventricles contract.
- Diastole is when the ventricles relax.

Topic questions

Describe the pathway of the electrical impulse for the heartbeat. **The signal for the heartbeat starts in the sinoatrial (S-A) node. It is then conducted through the atria and the atria contract. The signal is then delayed slightly at the atrioventricular (A-V) node. After another short delay, the signal is conducted into the ventricles and they contract.**

What is it called when the ventricles contract? **Systole.**

32.4 CIRCULATORY SYSTEM: BLOOD PRESSURE

- Blood pressure is the pressure that forms in the arteries due to blood being pumped into them.
- The systolic blood pressure is the "first number" of the blood pressure reading. It corresponds to the pressure in the arteries when the ventricles are contracting.
- The diastolic blood pressure is the "second number" of the blood pressure reading. It is the pressure in the arteries during ventricle relaxation.

Topic questions

Which number is listed first in a blood pressure reading, diastolic or systolic? **Systolic.**

If a person's blood pressure is 196/117, what do they have? **Hypertension.**

Why is the systolic blood pressure lower than the diastolic blood pressure? **I did not specifically state this, but I am hoping the student will be able to figure this out. The systolic pressure is higher because the ventricles are contracting. Since the ventricles are contracting, there is more blood being forced into the arteries. The more blood that is in the arteries, the higher the pressure is.**

32.5 BLOOD

- Blood is technically a connective tissue, but it is so important to the circulatory system that we discuss it here.
- Blood is made up of red blood cells, white blood cells, platelets, and plasma.

Topic questions

What are the components of blood? **Red blood cells, white blood cells, platelets, and plasma.**

What is plasma? **It is a liquid mixture containing proteins, ions, nutrients, and wastes in which the blood cells are suspended.**

What does a red blood cell look like? **A biconcave disc.**

What does a red blood cell contain that is so important? **Hemoglobin?**

What does this molecule do? **It carries oxygen to the tissues from the lungs and some carbon dioxide from the tissues to the lungs.**

What is oxyhemoglobin? **It is the oxygen-hemoglobin molecule complex; when oxygen binds to hemoglobin it is called oxyhemoglobin.**

32.6 RESPIRATORY SYSTEM: ANATOMY

- The respiratory system is divided into the upper respiratory and the lower respiratory tract.
- The anatomy is discussed.

Topic questions

What is the pathway that air takes into the lungs, staring with the nose? **The air passes through the nose, sinuses, pharynx, past the epiglottis, larynx, past the vocal cords, into the trachea, the bronchi, bronchioles and, finally, into the alveoli.**

What is an alveolus? **It is an air sac in the lung in which gas exchange occurs.**

32.7 RESPIRATORY SYSTEM: PHYSIOLOGY

- The alveolus is the functional unit of the lung.
- Respiration occurs in the alveolus.
- Inhalation is active, exhalation is passive.

Topic questions

Is inhalation active or passive? **Active.**

How does exhalation occur? **Through the elastic recoil of the lungs.**

What is ventilation? **It is the process of moving air in and out of the lungs.**

What is respiration? **It is the process of gas exchange.**

32.9 KEY CHAPTER POINTS

- The purpose of the circulatory system is to move oxygen, hormones, and nutrients to tissues/organs and carbon dioxide—as well as wastes—from tissues/organs.
- The circulatory system of humans and all vertebrates is a closed, two-loop system. It is composed of a heart, arteries, veins, and capillaries.
- The human heart has four chambers and four valves.
- The heart muscle is specialized to contract and conduct electricity.
- Blood pressure is a measure of the pressure in the arteries during the period of time the ventricles beat (systole) and the period of time they do not beat (diastole).
- Blood is a connective tissue and is made up of red blood cells, white blood cells, platelets, and plasma.
- The purpose of the respiratory system is to exchange gasses.
- The respiratory system is composed of the upper and lower respiratory tract.
- Gas exchange occurs through diffusion.

33 Human Anatomy and Physiology IV
Digestive, Musculoskeletal/Connective Tissue, and Integumentary Systems

33.0 CHAPTER PREVIEW

In this chapter we will:

- Discuss the anatomy of the digestive system.

- Investigate the physiology of the digestive system with special attention to:
  - salivary glands and amylase
  - peristalsis
  - the mechanical actions of the stomach and the enzyme pepsinogen
  - the function of the duodenum, jejunum, ileum, gall bladder, liver, pancreas, and colon
  - the absorptive capacity of villi and microvilli

- Discuss the structure and function of the three types of muscles—skeletal, cardiac, and smooth.

- Learn the structure and function of bone.

- Review the function of ligaments and tendons.

- Investigate joint types.

- Study the structure and function of the integumentary system (skin).

33.1 DIGESTIVE SYSTEM: GENERAL

- The human body needs to ingest proteins, fats, carbohydrates, minerals, vitamins, and water to be healthy.

- These materials are processed and absorbed by the digestive system.

Topic question

Our body can make all the vitamins needed to be healthy. True or False? **False. Our body can only make vitamin D. The rest need to be ingested and absorbed to remain healthy.**

33.2 DIGESTIVE SYSTEM: MOUTH

- The mouth mechanically and enzymatically begins to process food.

Topic questions

What do teeth do? **They mechanically break food into smaller pieces.**

What is amylase? **It is an enzyme secreted into the mouth that begins to chemically break starch into maltose.**

33.3 DIGESTIVE SYSTEM: ESOPHAGUS

- The esophagus propels a bolus of food into the stomach through waves of muscular contractions.

Topic questions

What is peristalsis? **Waves of muscular contractions that propel food through the GI tract.**

What is the esophageal sphincter and what does it do? **It is the muscular junction between the stomach and the esophagus that prevents food from moving into the esophagus from the stomach.**

33.4 DIGESTIVE SYSTEM: STOMACH

- The stomach further mechanically and chemically breaks food down. It prepares the food to enter the duodenum.

Topic questions

What is chyme? **It is the acidic paste mixture of food after it is churned and broken down by the stomach.**

How does pepsinogen work and what does it do? **Pepsinogen is an inactive enzyme. It is secreted into the stomach by stomach cells. When stomach acid is pumped into the stomach by stomach cells, pepsinogen is converted to pepsin. Pepsin is an enzyme that breaks down proteins.**

33.5 DIGESTIVE SYSTEM: INTESTINES, PANCREAS, AND LIVER

- The small intestine is made up of the duodenum, jejunum, and ileum.

- The large intestine is also called the colon

- The liver makes bile and stores it in the gall bladder.

- The gall bladder releases bile into the duodenum when food enters it.

- The absorptive area of the intestines is greatly increased by foldings called villi and microvilli.

Topic questions

What do the pancreas and liver empty into the duodenum and how do these chemical contribute to digestion? **The pancreas releases many enzymes into the duodenum to further break down fats, proteins, and carbohydrates. The liver makes bile and stores it in the gall bladder. The bile is released into the duodenum and helps the enzymes of the pancreas break down fats better.**

Once food leaves the duodenum, is the breakdown process of digestions completed? **Yes, once the food enters the jejunum, it is broken down into elemental forms of the fats, carbohydrates, and proteins.**

What are villi and microvilli and why are they important? **Villi are foldings of the inside of the small intestine, and microvilli are foldings of the villi. They are important because the foldings greatly increase the absorptive ability of the small intestines. All the nutrients, vitamins, and minerals needed are absorbed by the villi and microvilli.**

33.6 MUSCULOSKELETAL/CONNECTIVE TISSUE SYSTEM: GENERAL

- The function of the musculoskeletal system is to move and support the body.
- This system consists of muscles, tendons, ligaments, bones, and the extracellular matrix.

Topic questions

What function does the skeleton provide? **Support.**

What is the function of muscles? **Movement.**

What is the function of ligaments and tendons? **Ligaments hold bones together (connect bone to bone) and ligaments hold muscles to bones.**

33.7 MUSCULOSKELETAL/CONNECTIVE TISSUE SYSTEM: BONES

- The axial skeleton is the skull, spine, ribs, and sternum.
- The appendicular skeleton is the clavicles, pelvis, scapulae, arms, and legs.
- There are two types of bone—compact and spongy.

Topic questions

What is the periosteum? **It s the membrane covering all bones.**

What are the two types of bone? **Compact and spongy bone.**

What are Haversian canals? **They are canals through compact bone that allow blood vessels and nerves to pass through.**

Where is bone marrow located? **In the center of bone.**

Why is red bone marrow important? **Because it makes all the blood cells in the body.**

How do most bones form? **Through the process of cartilage first forming the structure of the bone, then bone cells moving into the area and then making bone. This process is called ossification.**

33.8 MUSCULOSKELETAL/CONNECTIVE TISSUE SYSTEM: LIGAMENTS AND JOINTS

- Joints form where two bones come together.
- There are several different types of joints, all classified based on the movement they allow.

Topic questions

What is a hinge joint; give an example? **A joint that move in one plane. An example is the knee or elbow.**

What is a ball and socket joint; give an example? **It is a joint that allows movement in all planes. Examples are the hip and shoulder joints.**

33.9 MUSCULOSKELETAL/CONNECTIVE TISSUE SYSTEM: MUSCLES AND TENDONS

- There are three types of muscle—skeletal, smooth, and cardiac.
- Muscles attach onto bones through tendons at attachment sites.

Topic questions

What is the difference between smooth and skeletal muscle? **There are several answers for this question. Smooth muscle is not under voluntary control; skeletal muscle is. Smooth muscle does not have striations; skeletal muscle does. Smooth muscle cells contain one nucleus per cell; skeletal muscle cells contain multiple nuclei per cell.**

What is the opposite joint movement of flexion? **Extension.**

33.10 MUSCULOSKELETAL/CONNECTIVE TISSUE SYSTEM: MUSCLE PHYSIOLOGY

- Muscles contract through the actions of actin and myosin after receiving an impulse from the motor nerve.
- Acetylcholine is the neuromuscular transmitter.

Topic questions

What is acetylcholine and what happens when it is released at the neuromuscular junction? **Acetylcholine is the neuromuscular transmitter. When the nerve releases it, it causes the muscle to contract.**

What is a sarcomere? **It is the functional unit of a muscle. During a muscle contraction, each sarcomere contracts, moving z-lines closer together.**

33.11 INTEGUMENTARY SYSTEM

- This system consists of the skin, nails, and hair.
- Skin is actually an organ.
- Skin is made up of two layers—the dermis and epidermis.

Topic questions

Skin is an organ. True or False? **True.**

What are the cells called which make up the skin? **Epithelial cells.**

What is the lower layer of the skin called? **The dermis.**

33.13 KEY CHAPTER POINTS

- The function of the digestive system is to break down and absorb the nutrients needed to sustain life and to eliminate wastes that are left over.

- The GI system has many specialized parts; these function to break down and absorb food.

- Food is prepared for absorption in the mouth, esophagus, stomach, and duodenum.

- Food is absorbed in the jejunum and ileum; water is absorbed in the colon.

- The function of the musculoskeletal system is to move and support the body.

- Bones support; ligaments attach bones to bones; tendons attach muscles to bones; muscles move.

- Muscles contract through the actions of actin and myosin.

- The integumentary system in the covering of the body. Its purpose is protection. It includes hair, nails, and skin.

34 Ecology

34.0 CHAPTER PREVIEW

In this chapter we will:

- Introduce basic concepts and terms of ecology.
- Investigate the dynamics of population studies.
- Discuss food chains, food webs, food pyramids, and trophic levels.
- Define the differences between mimicry and camouflage and study examples of both.
- Investigate the hydrologic, carbon, and oxygen cycles.
- Delve into the controversial topic of global warming.
- Discuss the seven land biomes, as well as marine and fresh water biomes.

34.1 OVERVIEW

- Ecology is the study of the relationships living organisms have with one another and their physical environment.
- Ecologists are the scientists who study ecology.

Topic question

What is the study of the relationships living organisms have with one another and their physical environment called? **Ecology.**

34.2 BIOSPHERE

- The biosphere represents the greatest level of complexity in ecology. It is the thin, physical layer of the earth in which all living organisms are found.

Topic question

What physical area of the earth does the biosphere cover? **It is about thirteen miles thick and extends from the deepest ocean depths (about seven miles below the sea level) to mountainous regions (about six miles above sea level).**

34.3 ECOSYSTEMS

- An ecosystem is defined as the association and interaction of all living organisms within their physical environment.

Topic question

What is an ecosystem? **The association and interaction of all living organisms within their physical environment.**

34.4 BIOTIC AND ABIOTIC MASS

- Biotic mass is all the living organisms in an ecosystem.
- Abiotic mass is all the nonliving things in an ecosystem.

Topic questions

Is a tree biotic or abiotic mass? **Biotic.**

What about water? **Abiotic.**

34.5 COMMUNITIES

- A community is all the organisms present and interacting in an ecosystem.

Topic question

If an ecosystem is defined by the researcher and includes trees, rocks, flowers, grass, sand, dirt, water, concrete, a wooden staircase, birds, insects, and fish, which of these are part of the ecosystem's community? **The trees, flowers, grass, birds, insects, and fish are all part of the community.**

34.6 POPULATIONS

- A population is defined as all the members of a certain species living at one place at one time.

Topic question

Which would you usually expect to contain a larger number of organisms in an ecosystem, a population or community? **Normally a community would be larger because a community includes all the organisms in an ecosystem and a population includes all the members of a certain species in an ecosystem.**

34.7 POPULATION DENSITY

- Population density is the number of organisms per given area of land.

Topic question

What information does population density provide? **An idea of how crowded the population is in an ecosystem.**

34.8 POPULATION DISPERSION

- Population dispersion is how spread out a population is.
- Populations disperse in three general ways: clumped, random, and even.

Topic question

If you were studying a population for their dispersion pattern and found that there were small concentrations of squirrel populations scattered here and there, what type of pattern would this be? **Clumped.**

34.9 GROWTH RATE

- The change of the population size is called the growth rate.

Topic questions

If you are studying an ecosystem and find that one of the populations has a negative growth rate, what does that mean? **The population is getting smaller. A negative growth rate indicates the population is not getting larger, but smaller.**

What factors affect growth rates? **Immigration, emigration, birth rates, and death rates.**

34.10 LIFE EXPECTANCY

- Life expectancy is the length of time an individual can be expected to live.

Topic question

You are studying a population of cardinals in an ecosystem. You have meticulously banded the birds for identification and followed their behaviors. You find that after twenty-five years of research, the average age of the birds when they die is six years old. There was one bird that lived to twelve. What is the life span and life expectancy of this population? **The life expectancy is the average length of time an organism can be expected to live, which is six years old in this example. The life span is the oldest known organism, which is twelve.**

34.11 ENERGY TRANSFER

- One of the important aspects of ecosystems is the transfer of energy from one population to another and one organism to another.

Topic question

34.12 The Food Chains

- The sun is the ultimate source of energy for almost every ecosystem.
- The relationship that one organism has to another in regards to transfer of energy is called the food web.

Topic questions

What is a consumer that eats a producer called? **A primary consumer.**

What is a consumer called that eats a secondary consumer? **A tertiary consumer.**

What is the transfer of energy between all organisms in an ecosystem called? **A food web.**

34.13 TROPHIC LEVELS AND FOOD PYRAMID

- A food pyramid is a tool to help visualize the energy available in an ecosystem and the transfer of energy from one organism in the food web to the next.
- A trophic level is the position that an organism occupies on the food pyramid.
- Only about 10% of the available energy from one trophic level is transferred to the next higher trophic level.

Topic questions

The organism on the bottom of the food pyramid has the highest or lowest amount of biomass and energy available to the ecosystem? **The highest.**

When energy is transferred from one trophic level to the next, how much energy is lost? **About 90%.**

34.14 ADAPTATIONS FOR BETTER SURVIVAL

- Organisms have a wide variety of adaptations to help them be better predators.
- Prey organisms have adaptations that make them better at avoiding being eaten.
- Camouflage and mimicry are two types of these adaptations.

Topic question

What is mimicry? **An adaptation in which a harmless species looks like a dangerous one.**

34.15 ABIOTIC COMPONENTS OF ECOSYSTEMS

- Important abiotic components of an ecosystem are carbon, oxygen, and water.
- The recycling of molecules in an ecosystem is called the biochemical cycle.

Topic question

What is the biochemical cycle? **It is the process of recycling various molecules within ecosystems.**

34.16 THE WATER CYCLE

- The water cycle is also called the hydrologic cycle. It describes the pathway water takes to get to the earth's surface from the atmosphere and then either return to the atmosphere or trickle into the earth to become ground water.

Topic questions

What are the usual pathways of the hydrologic cycle in a terrestrial ecosystem? **Water enters the ecosystem via precipitation. Water can then exit the ecosystem through ground water after it enters the earth through infiltration. After infiltration, it can be stored in the pores of rocks, soil, or aquifers. Water can also exit the system through evaporation, transpiration, or runoff.**

How does transpiration differ from evaporation? **Evaporation is water on the earth's surface turning into water vapor. Transpiration is the loss of water by plants into the atmosphere.**

What is a watershed? **It is an area where runoff water accumulates.**

34.17 THE CARBON AND OXYGEN CYCLES

- The air we breathe is 78% nitrogen, 21% oxygen, 0.935% argon, and 0.033% carbon dioxide.
- Carbon from the atmosphere is fixated into organic molecules by photosynthesis. During photosynthesis, oxygen is released and re-supplied into the atmosphere. Animals use the oxygen provided by photosynthesis for cellular respiration. During cellular respiration, carbon dioxide is released into the atmosphere, replenishing that molecule to be used during photosynthesis.

Topic questions

How is most of the atmospheric oxygen replenished? **Photosynthesis.**

How is atmospheric carbon dioxide replenished so it can be used for photosynthesis? **Cellular respiration and burning fossil fuels.**

34.18 THE GREENHOUSE EFFECT

- The greenhouse effect refers to a process by which the greenhouse gasses trap heat around the earth, just like the glass panes of a greenhouse trap heat.

Topic question

Is the greenhouse effect "bad." **No. Without the greenhouse effect, there would be no life on earth. The earth would be too cold.**

34.19 GLOBAL WARMING (?)

- The concept of global warming is that the earth is progressively becoming warmer because of the increasing greenhouse effect of carbon dioxide in the atmosphere. This is proposed as happening due to the increased concentration of carbon dioxide as a result of burning fossil fuels.
- The other side of the argument is that the temperatures on the earth have only been measured for the past one hundred years. How do we know that there is not a cyclical pattern of increases and decreases in the normal temperatures of the earth. There may be some indication that this is the case as there was likely a prolonged period of time several hundred years ago when the earth was considerably warmer than it is now.

Topic question

Do you think that global warming is occurring or do you need more information to make a certain determination?

34.20 BIOMES

- There are seven major terrestrial biomes on earth—tundra, taiga, temperate deciduous forest, temperate grassland, desert, savanna, and tropical rain forest.
- There are also aquatic biomes.

Topic questions

What is a biome? **A biome is a major regional group of distinctive plant and animal species in a common physical environment.**

How many terrestrial biomes are recognized on earth? **There are seven.**

What is a photic zone? **A photic zone is an area of an aquatic biome where light can penetrate and photosynthesis can occur.**

34.22 KEY CHAPTER POINTS

- Ecology is the study of the relationships of living things with one another and with their environment.

- Ecologists study ecosystems. An ecosystem is the association and interaction of all living organisms within a defined physical environment.

- Ecologists study individuals, populations, and communities.

- One of the important concepts of ecology is understanding the transfer of energy and food webs that exist in ecosystems.

- Trophic levels on the food pyramid lose about 90% of the available energy when moving from a lower trophic level to the next higher one.

- Organisms have a variety of adaptations to make them better predators or better at eluding predators.

- Ecologists are also interested in the cycles that various critical molecules of life follow, such as the water cycle, the carbon cycle, and the oxygen cycle.

- The greenhouse effect is the trapping of heat around the earth caused by several different gasses. It is a necessary process for life on earth to exist because without it, the earth would be too cold to support life.

- Carbon dioxide is a major component of the greenhouse gasses.

- The levels of carbon dioxide have been climbing in the past century; this is implicated by some scientists to be the cause of global warming.

- Global warming is not a scientific fact and further research needs to be conducted to ascertain whether it is happening or not.

- The land on earth is divided into seven biomes.

- Water ecosystems are included in marine or freshwater biomes.